# NATURE'S ELECTRICITY

# NATURE'S ELECTRICITY

## CHARLES K. ADAMS

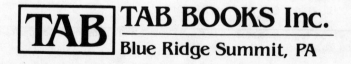

TAB BOOKS Inc.
Blue Ridge Summit, PA

FIRST EDITION

SECOND PRINTING

Copyright © 1987 by TAB BOOKS Inc.

Printed in the United States of America

Reproduction or publication of the content in any manner, without express permission of the publisher, is prohibited. No liability is assumed with respect to the use of the information herein.

Library of Congress Cataloging in Publication Data

Adams, Charles K.
    Nature's electricity.

    Includes index.
    1. Electricity—Popular works.   I. Title.
QC527.A32   1986                    537                    86-6010
ISBN 0-8306-2769-3 (pbk.)

TAB BOOKS Inc. offers software for sale. For information and a catalog, please contact TAB Software Department, Blue Ridge Summit, PA 17294-0850.

Questions regarding the content of this book
should be addressed to:

    Reader Inquiry Branch
    TAB BOOKS Inc.
    Blue Ridge Summit, PA 17294-0214

Cover photograph courtesy of National Oceanic and Atmospheric Administration (NOAA).

# Contents

**Introduction**                                                    vi

**1  Overview**                                                      1

Nature's Electrical Generators—Electron Flow and the Concept
of Charges—The Electroscope

**2  Piezoelectricity**                                              28

Solid Matter—Ions—Applications—Characteristics—Demonstrations—LCD

**3  Static Electricity**                                            57

History—The Leyden Jar—Static Charge Buildup on Plastic—The
Triboelectric Series—Generating a Charge in Air—Voltages Encountered in Static Electricity—Electrostatic Voltmeter—Detecting
the Presence of a Static Charge—Static Demonstrations—
Generating a Static Charge — Static Generators — The
Electrophorus—The Versorium—Electrostatic Motors—The Van
de Graaff Generator

**4  Atmospheric Electricity**                                       116

The Atmospheric Profile—Where Atmospheric Electricity Comes
From—Profile of a Thunderstorm—Measuring the Atmospheric
Charge—Lightning—Sferics—Using Atmospheric Electricity to
Drive a Static Motor

**Index**                                                            145

# Introduction

T HE PURPOSE OF THIS BOOK IS TO INTRODUCE THE READER TO some of the methods by which electricity is generated in nature. Several demonstrations and projects are included, which will familiarize the reader with the method under discussion. It also introduces the reader to electron current and how electrons flow.

Electricity can be generated in nature by various means. This book is limited to the discussion of piezoelectricity, static electricity, and atmospheric electricity. These are the most common types. Although in some cases the last two types may be partially related, they are covered separately because each is a major topic. Several experiments, or projects, are included to familiarize the reader with how it works, and how to demonstrate the results. These are all observable results, and the experiments range from simple demonstrations to the construction of equipment such as the Van de Graaff generator and static motors. The making of the equipment is not difficult and can be usually completed in a few hours or less. Some of the complex devices such as static generators will take longer.

The main object of the book is to stimulate thinking in this area. The better an individual understands nature, the better the individual can get along with nature. After all, Mother Nature is the major power, and to get along with her, we must understand her better. Mother Nature has many aspects, and this book covers only one of them. This book exposes a side of nature that is not often considered seriously when studying the action of nature. Some of the projects or experiments discussed here are of the parlor demonstration type: simple yet spectacular demonstrations. Others are serious and meaningful experiments, from which much can be learned. Some of the experiments, especially those dealing with static voltage generation, should be done carefully and treated as serious experiments—not toys.

The scope of this book is to cover these topics in a manner that does not require involved mathematics or a detailed understanding of physics or electronics. Most of the background knowledge that is required, such as the makeup of the atom and electron flow, is presented. After reading and studying this book, and working some, if not all, of the projects presented, it is hoped that the reader will find some aspect of this topic interesting enough to continue studies in this area. There are several additional experiments and developments that can be accomplished, and this work is not limited to the government- and university-sponsored research centers. The public libraries contain a great deal of information on these topics. Considerable work and many meaningful experiments can be conducted with a minimum of cost and equipment. The main requirements are interest and desire.

# Chapter 1

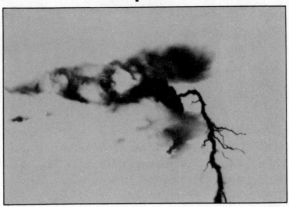

# Overview

**E** NERGY IS THE VERY ESSENCE OF LIFE, AND AT THE VERY FOUN-
dation of our existence. In the days when the earth was
young and forming, it was torn with violent, and cataclys-
mic releases of energy. In these violent times the single-cell
form of life appeared.

Energy makes life more comfortable for human beings.
It is the use of energy that is responsible for our ability to advance the standard of living. Energy comes in many forms:
solar energy, potential energy, nuclear energy, steam energy,
wood—and the list goes on and on.

- Solar energy is generated by the sun. It is about 1kW
per square meter, and is in the form of sunlight. Being sunlight, it is available only when the sun is shining. Solar energy
is directly converted to electricity using man-made solar cells.
Because the differences in temperatures and air pressures account for most of the wind, this is also considered a result
of solar energy.

- Potential energy is energy available due to the gravita-
tional position of the material with respect to the rest position. In other words, potential energy is that energy available
because of the action of gravity. For instance, in the precipitation loop when the moisture is in the clouds, it contains
considerable potential energy. The actual amount is a func-

tion of the mass of the moisture times the altitude of the water vapor. When the moisture condenses into rain and falls, some of the potential energy is released. The remaining potential energy is a function of the height above sea level. One of the most common examples of man's use of potential energy is harnessing a river flowing downhill. By damming it up, the water can be released in a controlled manner to turn generators. The potential energy of the river is the energy available when the water flows downhill.

- Nuclear energy is the natural radiation of the radioactive materials degenerating. The normal end result is heat. Many natural things contain measurable amounts of radiation. For instance, coal contains some radiation, and this is released into the atmosphere when the coal is burned.

- Steam energy is the energy available because of the temperature of a liquid. Typically, this is the energy of the steam after water has been vaporized. The required heat may be generated in several ways, including contact with lava. This steam energy is expanded in nature by such things as geysers and geothermal wells.

These are some examples of the many types of energy generated or released by nature. That is in addition to electrical energy, or electricity. Although electrical energy is the major energy form used by people, it is not the major form generated in nature. In many instances, electrical energy is the result of the conversion of another form of energy.

Electricity is used by everyone every day. The day starts with the electric alarm clock, the electric coffee maker makes the coffee, the electric lights illuminate the darkness, the TV provides entertainment, the air conditioner cools the house, electrical thermostats control the furnace that heats the home and buildings, the dishwasher washes the dishes, and the list goes on and on.

Electricity is responsible for most of the industrial expansion, because it powers the production lines that turn out the products and provide jobs for millions. We, as humans, like to think that we developed electricity, and that we alone can produce it. But Mother Nature has beaten us to the punch, so to speak, for Mother Nature has been producing electricity for millions of years. In fact, some theories state that electricity was the catalyst for the formation of life.

Electricity is the primary energy used in this country, and most of the industrialized countries of the world. That is when the fossil fuels used to generate electricity are not taken into consideration. Our modern society generates electricity using several methods, some of which produce considerable pollution. The commercial methods used to generate electricity are: plants powered by fossil fuels, hydroelectric dams, nuclear power plants, and windmills. Of these, the hydroelectric dams and the windmills are considered renewable energy sources. A renewable energy source is a source that will not be depleted when used, but keeps renewing itself. It does not require a fuel. Fossil fuels and nuclear power plants are nonrenewable energy sources.

Nature releases electricity in some colorful, violent, and noisy displays. The results of nature generating electricity are seen quite often, the most prevalent being lightning. The power of lightning can be observed when it strikes a tree or unprotected building. It has been known to shatter trees, melt sand, melt metal poles, start fires, and drive instruments crazy. It has followed a wire into a house and, in the form of a ball, dance around the room. There is at least one instance where all the metal nails were blown out of the roof of a house by lightning. There is another instance wherein a man, blind for years, regained his sight after being struck by lightning.

One of the more passive displays of nature's electricity is the northern lights, or aurora borealis. This colorful display of dancing lights is a result of cosmic particles interacting with the atmosphere. This particles enter the atmosphere and excite the oxygen atoms. A reddish glow is emitted when the atoms return to the normal energy state. During periods of high solar flare activity, these lights cover a broad area. Normally they occur over the North Pole. A less publicized display is the lights seen during some earthquakes, the most famous of these being in Japan. Another spectacular passive display is the so-called 'spook' or 'ghost' lights, balls of light that dance along the ground.

The shock from static electricity is another indication of nature's electricity. We have all experienced the shock after walking across some types of rugs or sliding across the car seat and touching a metal door handle. This is an aggravating result of the ability to generate static electricity. The small spark that is discharged is actually a small lightning discharge. In a larger sense, static electricity can start fires, cause explosions, and do all sorts of damage to electronic equipment.

## NATURE'S ELECTRICAL GENERATORS

Nature generates electricity in a number of ways, some of which are unique to nature, and some of which have been used by man. Following are some of the methods nature uses to generate electricity:

- Piezoeletricity
- Static electricity
- Atmospheric electricity
- Cosmic rays
- Acid batteries
- Salt water electrolysis
- Dissimilar metals
- Thermoelectricity

This book covers the first three methods, piezoelectricity, static electricity, and atmospheric electricity. Let us, however, first take a brief look at the others.

Cosmic rays are higher energy particles that permeate the universe. A majority of those that strike our atmosphere originate at the sun. These cosmic rays, when striking the upper atmosphere, release electrons and are responsible for such phenomena as the northern lights. Cosmic rays striking the moon induce an electrical charge on the surface. This causes dust particles to levitate from the surface, as seen in some of the moon photographs.

Natural acid batteries occur when an acidic liquid contacts some metals. One well-known natural battery is the citrus fruit battery. Citrus fruit contains an acid—citric acid—that will provide battery action when electrodes are inserted into the fruit. The acid provides the electrolyte for the battery. There are cases where acidic water, trapped in high mineral content soil or rocks, has produced the battery effect.

Salt water electrolysis is the action of the salts in salt water on some metals. The chemical reaction between the salt and some metals results in the release of free electrons, and therefore in current flow. This type of chemical reaction is also present where several types of corrosion occur.

4

When two dissimilar metals are joined, a small electrical potential (or voltage) is generated. Each metal has a chemical property known as the electromotive potential. When two metals whose electromotive potentials are significantly different are joined, a small electrical voltage is generated across the junction. This voltage is essentially equal to the difference in electromotive potential. This difference in electromotive potential can eat away the metals.

Thermoelectricity is the ability of some materials to provide electron flow when one side is heated with respect to the opposite side. It is a reversible process in which the application of a voltage across the material will cause one side to become warm and the opposite side to become cool. The understanding of this phenomena is an outgrowth of research into semiconductor theory and the development of advanced electronics. The explanation of why the thermoelectric materials work is beyond the scope of this book. Although man-made materials are available with demonstrated significant thermoelectric properties, it is a matter of conjecture whether the thermoelectric phenomena exists in nature. These man-made materials are used primarily in the reverse mode, where it provides heating or cooling when current is supplied.

## ELECTRON FLOW AND THE CONCEPT OF CHARGES

Before proceeding, let us look at the process of electron flow and free electron generation. Electrical current is the flow of electrons through some type of conductor. This requires a source of electrons that are free to move. Originally, before the concept of electrons and electron flow was developed, current was thought to flow from a positive potential to a negative potential. This concept is called classical current flow, where the current flows from a positive voltage source to a negative voltage source. It should be noted that for the purposes of this book, voltage and potential have the same meaning, and are used interchangeably.

The classical model of the atom consists of electrons, neutrons, and protons. The electrons contain the negative charge, the protons the positive charge, and neutrons make up most of the mass. Although this model of the atom is an oversimplification, especially considering modern nuclear physics, it is sufficient for our use.

Because the electron has a negative charge, electron flow is from the negative voltage source to the positive voltage source, the opposite that of classical current flow. This is illustrated by Fig. 1-1, which shows current flow in a simple battery circuit. Classical current flow and electron flow are shown. Notice that classical current flow is in the direction of arrow 1. This flow is from the positive terminal of the battery, through the resistor and bulb, to the negative battery terminal. Electron flow, as illustrated by arrow 2, is in the opposite direction. It is from the negative battery terminal, through the lamp and resistor to the positive battery terminal. In this book we will use electron flow as current flow.

To achieve electron flow requires a source of excess electrons (or a source of the absence of electrons) and a conductor to carry the current. For a conductor to carry current requires that some of the electrons in the atoms of the material be relatively easy to replace with other electrons. Electron flow consists of the movement of electrons down a conductor. Electrons enter one end of the conductor, and replace electrons in the atoms in the conductor. These atoms have an extra electron, and consequently, an electron is released to move on to another atom. The electrons that come out of the other end of the conductor are not the same electrons that entered the conductor originally.

Let us now look at the atom, which can be considered as the basic building block of material. It is the smallest division in which the material maintains its identity. Figure 1-2 shows the atomic diagram for a simple atom. As referenced

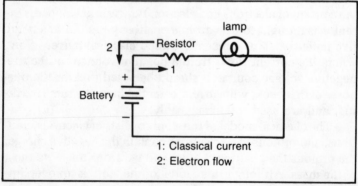

Fig. 1-1. Current flow in a simple battery circuit, illustrating both classical current flow and electron flow.

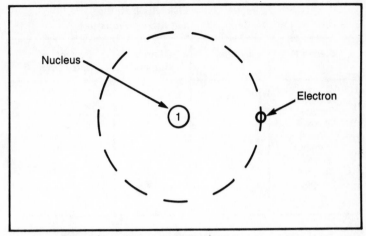

*Fig. 1-2. Simple atomic diagram for a simple atom.*

above, this is the classical (or Bohr's) model of the atom which will serve for our purposes. The electron orbits around the nucleus, or center part of the atom. The nucleus consists of protons and neutrons. The electron has one negative electrical charge. In fact, this is one of the definitions of electrical charge. Each proton has one positive charge that is equal to the electron charge, but the opposite polarity. The neutron has no charge, but makes up the mass of the atom. Because the atom has a total neutral charge, the number of protons in the nucleus equals the number of electrons in orbit around the nucleus. The positive and negative charges cancel each other, and the normal atom has no net charge.

The atom in Fig. 1-2 is an atom of hydrogen. It has one electron in orbit around the nucleus, which contains one proton and one neutron. This is the simplest of all the chemical elements, with an atomic number of 1 and an atomic weight of 1.008. All the chemical elements are represented in the periodic table of the elements. This periodic table lists all the known elements, with their atomic numbers and atomic weights. Also shown in the periodic table is the chemical symbol, the number of different orbits, and the number of electrons in each orbit. Table 1-1 lists the properties of some of the elements.

Notice that there is a definite maximum number of electrons in each orbit around the nucleus, and when one orbit is filled, the next orbit is started. The atomic weight of an ele-

**Table 1-1. Some Elements from the Periodic Table and Their Atomic Properties.**

| Element | Symbol | Atomic Number | Electronics in Outer Orbit |
|---------|--------|---------------|---------------------------|
| Hydrogen | H | 1 | 1 |
| Carbon | C | 6 | 4 |
| Oxygen | O | 8 | 6 |
| Sodium | Na | 11 | 1 |
| Aluminum | Al | 13 | 3 |
| Silicon | Si | 14 | 4 |
| Chlorine | Cl | 17 | 7 |
| Copper | Cu | 29 | 1 |

ment is defined as a number that indicates how many times as heavy one of the atoms of the element is as compared to 1/16 of the weight of the oxygen atom. In other words, oxygen is assigned the atomic weight of 16, and the other atomic weights are determined from that. The atomic number is the total number of electrons in orbit, or the number of protons.

Atomic physicists are delving deeper into the makeup of the atom, and the forces that bind it together. They are finding minute and exotic particles that are generated within the atom, or exist within the atom. The discussion of these is interesting and involved, and is beyond the scope of this book. Such a discussion requires a knowledge of atomic physics, complete with detailed mathematics.

Let us look at a more complex atom: aluminum. From Table 1-1, we see that aluminum has an atomic number of 13, with 13 neutrons and 13 electrons. The 13 electrons in orbit are divided between three orbits, with two in the innermost orbit, eight in the center orbit, and three in the outer orbit. The atomic diagram for aluminum is shown in Fig. 1-3. The number in the center is the atomic number for the material. This is a complex diagram, and can be confusing when drawn in this manner. Besides, the electrons in the outer orbit determine the majority of the characteristics of the atom. It is these electrons in the outer orbit that are available for electron flow, and determine whether the element is a conductor, insulator, or semiconductor. Although it is possible to free one of the electrons in one of the inner orbits, it takes a considerable amount of energy and is not normally accomplished.

The maximum number of electrons in each orbit is fixed, and their number depends on which orbit they are in. For every orbit except the first, the maximum number is eight. If it is an inner orbit, it may be 18 or 32. For example, the fourth orbit has a maximum of eight electrons if it is the outer orbit. If there are five orbits, the fourth orbit can have 18, and if there are six orbits of electrons, the fourth orbit can have 32. This gets complicated, and the reason is not thoroughly understood.

So, a simplified method of presenting complex atoms is to show only the electrons in the outer orbit. The inner circle represents the nucleus and the inner orbits. This is illustrated in Fig. 1-4, which is the simplified diagram for the aluminum atom. Notice that only three electrons are shown. These are the electrons in the outer orbit of Fig. 1-3. The number in the nucleus is three. Compare Fig. 1-4 with Fig. 1-3, and notice the simplification.

The number in the nucleus of Figs. 1-2 and 1-3 is the number of positive charges in the nucleus. Because the overall atom has a neutral charge, and the electrons have a negative charge, the positive charges in the nucleus must equal

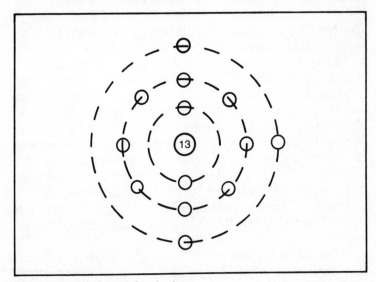

Fig. 1-3. Atomic diagram for aluminum.

9

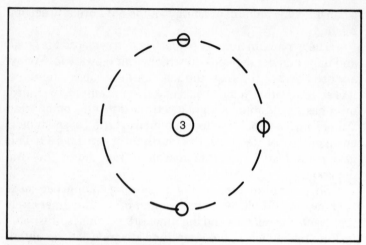

Fig. 1-4. Simplified atomic diagram for aluminum.

the number of electrons in orbit around the nucleus. The number in the center is not the valence of the atom. The valence of an atom is the number of electrons that it would take to fill the outer orbit. The valence for aluminum, as shown in Fig. 1-3, is five, because it would take five electrons to fill the outer orbit. The number in the center of the simplified diagram is, in reality, the maximum number of electrons in the outer orbit, minus the valence. It is also the number of electrons in the outer orbit. This concept is so important that it is presented again from a slightly different aspect.

Referring back to Fig. 1-4, the number in the center of this simplified atomic diagram is the resultant charge of the nucleus and all the electrons in the inner orbits. In other words, the nucleus and all the inner orbits of Fig. 1-3 are represented by the nucleus in Fig. 1-4. They are equivalent, but do not show the same detail. The charge of the nucleus is positive and the electrons have a negative charge, so the number of electrons in the inner orbits are subtracted from the original nucleus charge. Another way of determining this number is to visualize the atom as an atom with only the outer orbit of electrons, and that the nucleus consists of all the inner orbits of electrons and the nucleus.

The concept of electron flow in a solid might be difficult to comprehend. After all, the idea of a solid (the electron) moving through another solid (the atom) is against all logic. But an atom is made up of mostly spaces. The volume of the nucleus is only a very small percentage of the total volume of the atom. Most of the volume is taken up by the orbits of the electrons. The size of the electron is very small, even when compared to the nucleus, so there is plenty of room for the electron to move within the atom.

In aluminum, to achieve electron flow requires a source of electrons (a battery or voltage source) and a conductor (aluminum). Refer to Fig. 1-5, which illustrates electron flow in aluminum. The voltage applied across the conductor results in an electrical field. An electric field exists between two points of different electric potential. This electric field adds energy to the electron because opposite charges attract each other and like charges repel. This results in an attraction toward the positive charge, (right end of Fig. 1-5) and a repulsion from the negative charge (left end).

An aluminum atom has three electrons in the outer orbit, as discussed before. But, for an outer electron orbit to be full and not able to accept additional electrons requires that there are eight electrons in the outer orbit. This means that the aluminum atom will try to share electrons with five other atoms, but aluminum has only three electrons to share. Figure 1-5 gives a simplistic illustration of the results.

When an electron enters the material from the negative source at the left side, it joins the orbit of one of the atoms, as shown by track 1. This adds an extra electron to the atom, giving it a net negative charge. This charge, and the applied electric field result in an electron leaving that atom to join another atom (line 2). The electrons, when they are not captured by an atom, are moved toward the positive source. This results in an extra electron, or negative charge. This negative charge, added to the charge of the electric field, results in another (or the same) electron being expelled from the atom (3). This free electron is moved toward the right end of the conductor until it is captured by another atom, and the process repeats itself.

This process is repeated hundreds of times as one electron moves through a short conductor. Hundreds of electrons enter at one end and leave at the other end, forming hundreds of parallel paths through the material.

Another way to visualize electron flow through a wire is to compare it to a lake with a dam and a spillway. The water going over the spillway is not the same water that just en-

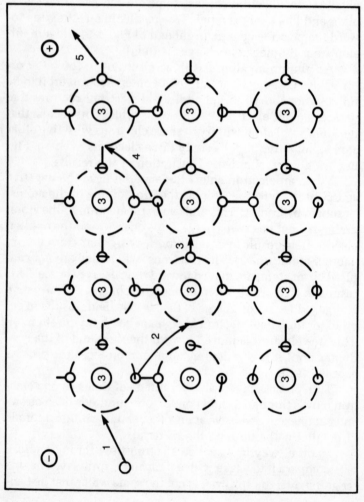

Fig. 1-5. Conduction in aluminum, showing electron movement.

tered the lake. The water that is flowing into the lake mixes with the rest of the water and might never make it over the spillway, especially if the water level drops. The water going over the spillway has been in the lake a long time, and has worked its way from the inlet to the spillway. The water entering the lake displaces water in the lake, causing the level to raise, and forcing water out.

The concept of electrical charge must be understood, because the term charge is used throughout this book. A charge can be considered as an accumulation of extra positive or negative charges. A charge can be either positive or negative, depending on the polarity. Polarity is either positive or negative, depending whether there are extra electrons or an absence of electrons to the basic neutral atom. A neutral atom contains the same number of electrons as protons. An atom with an excess of electrons has a net negative charge equal to the number of added electrons. The added electrons are negative, so this atom is called a negative ion. Figure 1-6 shows an atom with an extra electron added. This extra electron may be added in several ways, for example, by being in an electric field.

The atom, as shown, is any material with four electrons in the outer orbit, such as carbon or silicon. This is one of the disadvantages of using the simplified diagrams to illustrate the atom. The atomic number, and consequently, the

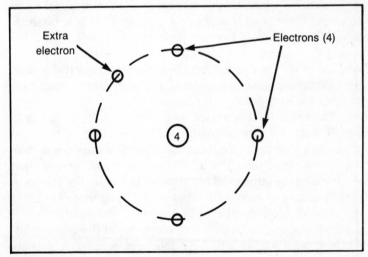

Fig. 1-6. Formation of a negative ion.

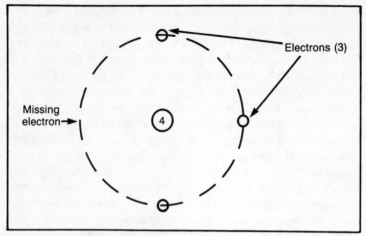

*Fig. 1-7. Formation of a positive ion.*

name of the material, is not readily distinguishable. An atom with one or more electrons removed has a net positive charge because the removed electron removes one negative charge from a neutral atom. This is called a positive ion, and is illustrated in Fig. 1-7. Notice that for both the negative and positive ions, it is the electron that is either added or removed. This is because, short of high-energy nuclear surgery, it is impossible to remove or add a proton to the nucleus. So this leaves only the electrons.

When the positive ion is created (by the removal of an electron) there is an absence of an electron. This is often referred to as a "hole," because there is a place or hole created where the electron is removed. So, when a figure is shown with an area of positive charge or an area of negative charge, the charges are generated by an excess or absence of electrons, or negative and positive ions.

The original atoms used to generate the ions shown in Figs. 1-6 and 1-7 both originally had four electrons in the outer orbit. Ions can be made from atoms with more or less than four electrons in the outer orbit, but atoms with one or eight electrons in the outer orbit are difficult to ionize. For instance, aluminum, with an atomic number of 13 and three electrons in the outer orbit, can be made a positive or negative ion. A positive ion will have only two electrons in the outer orbit, while a negative ion will have four electrons. The polarity of the ion is determined by whether the atom gained or lost

14

an electron, and not how many electrons are in the outer orbit.

Most atoms appearing in nature are neutral atoms, having neither a positive or negative net charge. To ionize an atom requires the addition of energy—to remove or add an electron. The ions can gain or lose more than one electron at a time, but this requires additional energy.

Sometimes positive or negative ions are shown only by the + (plus) sign for positive charges and the – (minus) sign for negative charges. One of the basic concepts of electricity is that unlike charges attract and like charges repel. The derivation of this concept involves mathematics and physics, and is beyond the scope of this book, but the concept simply means that a positive charge will attract a negative charge and repel a positive charge. Conversely, a negative charge will attract a positive charge and repel a negative charge. So when like charges are in close proximity, they repel each other with a measurable force. This is illustrated in Fig. 1-8, which shows that if two movable needles are charged the same polarity, they will tend to move away from each other. This is in a manner very similar to the way magnets behave.

Conversely, if the two needles are charged the opposite polarity and are brought in close proximity, they attract each other as illustrated in Fig. 1-9. The force of attraction or repulsion is evident where the charges are mobile and not fixed—where the material containing the charges is free to move. For instance, a needle suspended by a string is free to move, so is a pith ball, or a small piece of styrofoam suspended by a string. Under the proper conditions these can be made to move under the influence of electrical charges. When a comb is run through dry hair, the hair seems to be attracted to the comb and will "reach" out toward it. The hair generates a

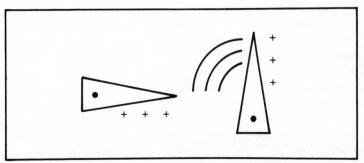

Fig. 1-8. Like charges on two needles repelling each other.

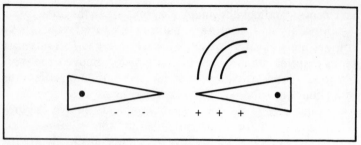

*Fig. 1-9. Opposite charges on two needles attracting each other.*

charge on the comb. (This is covered in more detail in the chapter on static electricity.)

When we talk about charges being attracted or repelled within a material, it is not the actual atoms that move in a solid. Although the ionized atoms might physically move in a gas or liquid, it is impossible for the atom to physically move in a solid. Realize that the ion, as discussed in this book, is an atom plus or minus an electron. For this atom to physically move requires at least partial disintegration of the solid. It is the electrons that move, and give the appearance that the ionized atom is actually moving. In Fig. 1-10, the upper diagram shows an even distribution of neutral atoms in a solid. These atoms are not ionized because there is no external energy applied. The circles without a polarity indicator represent neutral atoms.

If a positive charge is moved into close proximity to the left end, as shown in the lower diagram of Fig. 1-10, ions will be formed. The positive charges will attract electrons toward the left end, resulting in a negative charge at this end. The extra electrons forming the negative ions came from atoms near the right end. Consequently, these atoms become positive ions.

When the external charge is removed, the excess electrons will move toward the positive ions within the material. The electrons will combine with these positive ions, resulting in a neutral charge, as was the original case.

If, instead of a positive charge, a negative charge is moved into close proximity of the left end, the left end of the bar will become positively charged. This is because electrons will be repelled from the left end toward the right end by the negative charge. Therefore, the left end will be positively charged and the right end negatively charged.

A simplified explanation for the difference between a conductor and a non-conductor is as follows. The electrons in the outer orbit of the atom determine most of the properties of the material. When atoms combine they join by sharing the electrons in the outer orbit. If the outer orbit is filled, the material will not combine and will be what is called an ''inert gas.'' It is a gas because the atoms will not combine with each other as required to form a solid.

For a material to be a conductor requires that electrons be relatively free to move. The energy required to free an electron for movement is a measure of the electrical losses in the conductor. If the outer orbit of an atom is not filled with electrons that are shared, then the energy required to free those not shared is less than to free the shared electrons.

Figure 1-11 illustrates the concept of shared electrons. The electrons in the outer orbits of the atoms are shared by more than one nucleus. This means that the shared electrons have more force holding them in place, so it takes more energy to free them for electron flow. Remember that electron flow through a conductor is started by electrons entering the conductor at the negative end. These electrons are captured by atoms. Because these atoms have an excess of electrons, an electron is freed to be captured by other atoms. This continues until electrons come out of the other end. Another way of stat-

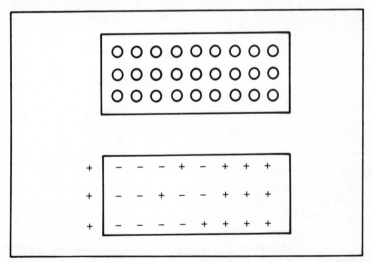

Fig. 1-10. Movement of charges within a solid. Upper shows uniform negative charge. Lower shows influence of positive external charge.

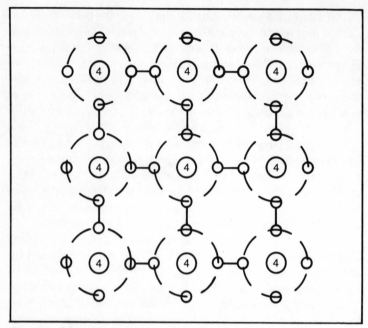

Fig. 1-11. The concept of shared electrons.

ing it is that electrons flow into the negative end of the material and move toward the positive end.

During the transition the electrons will collide with atoms and displace electrons in the outer orbits of these atoms. These electrons will, in turn, move toward the positive end until they either flow out the material, or collide with another atom and displace another electron. This process repeats itself until the electrons flow out of the end, and the process continues as long as current flows.

## THE ELECTROSCOPE

One of the simplest devices used for determining if an electrical charge is present is the electroscope. It works on the principle that like charges repel. This is a transparent device with two thin plates suspended from a metal hanger that is connected to a metal knob on the top of the device.

Figure 1-12 show the cross section of an electroscope. When an electrical charge is introduced to the metal knob (1) on the top, this charge is carried down the metal hanger (2) to the two thin metal plates, or leaves (3) suspended from the metal hanger. Because the charge is the same polarity on both plates or leaves, the plates will separate at the bottom (as shown). The container (4) is transparent and usually made from glass so the movement of the plates can be observed. The use of a clear plastic container is not advisable because of the ease with which most plastics pick up a static charge. The insulator (5) supports the metal hanger and insulates the glass container from the metal knob and hanger. If the charge is conducted to the container, it will interfere with the operation. In addition, the charge is not dissipated into the container. When there is no charge on the plates, or when the charge has dissipated, the plates will hang down and together.

There are two ways to use an electroscope: induction and conduction. Figure 1-13 shows an electroscope being charged by induction. The charge will remain in the electroscope until it is removed or it dissipates. When charging the electroscope by induction, the charge is not touched to the electroscope.

Referring to Fig. 1-13, when a negatively charged object is placed close to the metal knob of the electroscope, and the

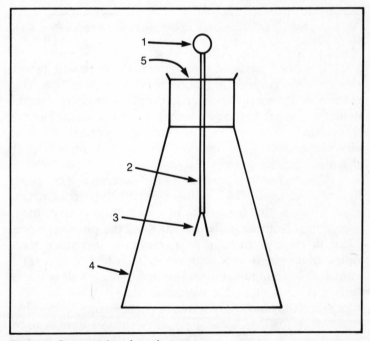

*Fig. 1-12. Cross section of an electroscope.*

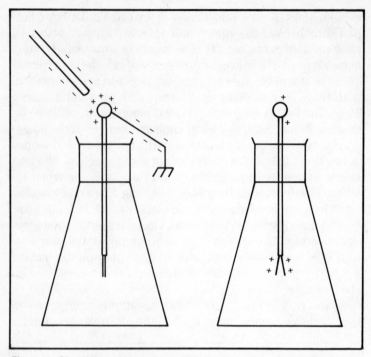

*Fig. 1-13. Charging an electroscope by induction using a negatively charged object.*

knob grounded, negative charges from the knob will be conducted to the ground, and away from the electroscope. This is because like charges repel, and the negative charges on the object will repel the negative charges on the knob. Remove the ground, then remove the negatively charged object. A simple means of grounding the electroscope knob is to touch it with a finger.

When in use, it is possible for the electroscope container to become charged. This will interfere with the operation. One way to reduce this possibility is to be careful of the surface where the electroscope is placed. Keep the physical movement of the electroscope to a minimum to reduce static buildup. If there is a problem, wrap the lower portion of the container in aluminum foil and ground the foil. This will neutralize any charge on the container.

When the object is removed, the net charge on the electroscope is positive, because negative charges have been removed in the above step. This leaves the system charged positive, so they will separate as shown in the figure.

Figure 1-14 shows charging the electroscope by conduction. When a negatively charged object touches the electroscope, the negative charges from the knob are again repelled. This time, these charges go through the metal hanger to the electroscope leaves, leaving a net negative charge on the leaves. The leaves will separate. When the charged object is removed, the leaves tend to go together.

The two examples above will behave in the same manner if a positively charged object is used instead of a negatively charged object. This makes sense because the electroscope works on the principle of like charges repelling to force the leaves apart, and not on the polarity of the charge. The charge is introduced into the electroscope at the knob. Because the knob, leaf hanger, and leaves are conductive, the charge will travel down the hanger to the leaves. If both leaves are making good contact with the hanger, the charge will go to both leaves. The leaves are charged with the same polarity, and like charges repel, so the leaves will be forced apart.

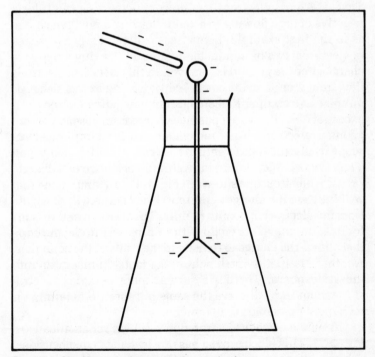

*Fig. 1-14. Charging an electroscope by conduction.*

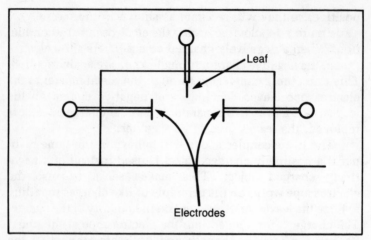

*Fig. 1-15. A twin electrode electroscope ready for operation.*

The distance the leaves are separated is an indication of the strength of the charge.

Ground is used as an infinite source or sink of electrons. That is, excess electrons can be conducted to ground for a negative charge flowing to ground, leaving a positive charge as in the first example. Likewise, positive charges (absence of electrons) can be conducted to ground, leaving a negative charge after the ground is removed. In this last case, electrons flow from ground to the electroscope, leaving excess electrons on the electroscope. Remember that a negative charge is an excess of electrons and a positive charge is an absence (or not enough) electrons. Also remember that the simple electroscope itself cannot differentiate between a positive charge and a negative charge, it only indicates the presence of a charge.

The electroscope shown in Fig. 1-12 is a simple one that will indicate if a charge is present. The charge will dissipate from the electroscope quite rapidly. There are several reasons for this, but the major factor is the air inside the electroscope that allows the charge to transfer to the walls of the container. Another dissipation path is from the metal knob on top into the atmosphere. There are several more complex electroscopes, but they all serve the same purpose: to determine if a charge is present.

A twin electrode electroscope is shown schematically in Fig. 1-15. This electroscope has one leaf and two electrodes. A high voltage is applied across the electrodes, then a charge

22

is introduced to the leaf. The leaf will move toward the electrode of the opposite polarity as that of the charge, so the polarity of the charge can be determined.

As the first piece of equipment, let us construct a simple electroscope. This electroscope will be used in several of the projects in the following chapters, so it is constructed here in the introduction. An introduction of how to use it is also included.

A photo of the electroscope is given in Fig. 1-16, and the plans are shown in Fig. 1-17. The parts required are:

- An empty baby food jar, or jar of similar size.
- Brass knob (drawer pull) about 3/4 inches across.
- Cork stopper between 3/4 and 1 inch in diameter.
- Bolt for drawer pull long enough to go through the stopper.
- Nut for above bolt.
- Paper clip or similar size solid wire.
- Two small pieces of aluminum foil.

An empty baby food jar is used because it is glass with

Fig. 1-16. Photo of a "baby food jar" electroscope.

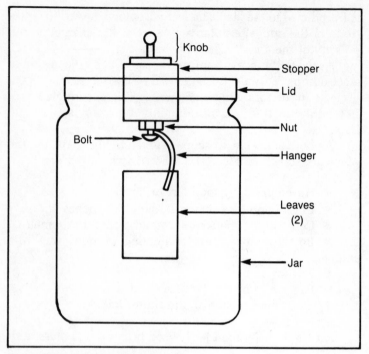

*Fig. 1-17. Plans for the electroscope.*

a metal lid, and is the correct size. If one is not available, any glass jar about 2 inches in diameter and 3 inches high, with a metal lid, can be used. Make sure it is clean.

The brass knob is a small drawer pull about 3/4 inch in diameter. This is available from most hardware stores. Most of them have a coating of clear acrylic that must be removed from the top portion. Use fine emery paper to remove the coating, because it will not work if the coating is left on. The knob must be solid metal in order to be able to conduct the charge to the leaves.

A cork stopper supports the knob and the hanger for the leaves. Use a sharp knife or razor blade to cut the stopper to about 1 inch in length.

Cork is used because it must be a good insulator and a non-conductor for static charges. Several of the rubber stoppers and plastic caps will conduct static charges, thus dissipating the charge before it gets to the leaves. If a cork stopper is not available, a 1-inch piece of wooden dowel about 3/4 inch in diameter can be used. Taper the dowel slightly with sandpaper. The paper clip is used as the hanger for the leaves, and the small pieces of aluminum foil are the leaves. Examine the paper clip carefully to determine if it is coated with some material. Some of them are coated with a clear material, and this must be removed to assure good electrical connection.

Drill a hole in the jar lid large enough for the stopper to fit part way down in. Remove all sharp projections from the metal lid with a file. Drill a hole through the stopper for the bolt. Next, assemble the lid, knob, hanger assembly as shown in Fig. 1-18. Cut and bend the paper clip as shown. Leave one of the loops intact, and bend the other around the bolt shank, just above the head. Then run the nut down on the bolt and tighten it to secure the paper clip hanger to the bolt. Push the bolt through the stopper, which is pressed into the lid. Attach and tighten the knob. This clamps the stopper in the lid. Bend the paper clip until the center of the loop is below the head of the bolt. When assembled, the hanger is secured to the bolt, which is secured to the knob. All the parts must be tight, and the stopper secured tightly in the hole in the lid. The loop from the paper clip must be positioned directly below the bolt with the rounded portion of the loop at the bottom.

Next, cut out the leaves from the aluminum foil. Make two of them the same size with a hole near the top, as shown. Press the pieces of foil flat with a small wooden block if re-

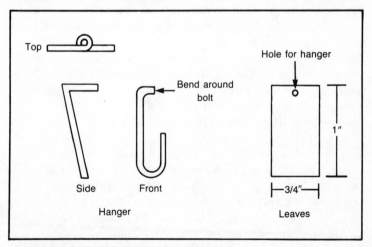

*Fig. 1-18. Details of the lid, knob, and hanger assembly.*

25

quired. The leaves must be flat and hang together, and not be bound together or to the paper clip. Slip the holes in the leaves over the paper clip so that they hang at the lowest portion and are free to move.

Attach the lid of the jar, complete with the knob and leaves, to the jar. The knob should project from the top, and the leaves should hang freely inside the jar, as shown. The leaves must not touch the bottom of the jar. Shake the jar slightly to make sure that the leaves are free to move. The electroscope is now assembled and ready for use.

To test the electroscope requires charging it with some static electricity. There are several ways of doing this. One of the easiest ways is to use a comb. Most combs will work—hard rubber is one of the best. If the hair is dry, run the comb through your hair a few times and bring it close to the knob of the electroscope. The hair must be thoroughly dry for the comb to pick up a static charge. If this is impossible, try rubbing the comb with wool, felt, silk, or similar material. Alternatively, rub a piece of styrofoam with a wool blanket. Yet another alternative is to rub a glass or Pyrex glass, rod, or tube (such as a test tube) with a wool blanket. Whichever method is used, determine if an electrical charge has been generated by trying to pick up small bits of paper with the charged object.

Notice that most of the materials mentioned to generate charges consist of one flexible object (wool blanket) and one rigid piece (a piece of styrofoam). To generate an electrical charge, hold the flexible material in one hand and the rigid object by one end in the other hand. Place the rigid object on top of the flexible object, and tightly wrap the flexible object around the rigid object. Move the rigid object back and forth several times, in relation to the flexible object. The tighter the flexible object is held against the rigid object, the better the charge generation, so hold the flexible object around the rigid object tightly enough so that there is considerable friction, but not tightly enough to damage either object.

If there is a static charge generated, and if the electroscope is working properly, the leaves will spread apart when the comb is brought close to the knob. Touch the knob with the comb, and at the same time touch the knob with a finger of the other hand. Remove the finger from the knob and the leaves should be back together. Remove the comb and the leaves should separate. They will stay separated for a couple seconds until the charge dissipates.

The total procedure—touching the comb to the electroscope to removing the comb—should not take over four seconds. This is because in a normal atmosphere the charge will dissipate into the air.

If the electroscope did not work, there are two possible reasons: The most probable is that there was no charge generated by the comb. Take a dry styrofoam coffee cup and rub it with a piece of felt or nylon, then touch the styrofoam to the knob of the electroscope. The leaves should separate. If they do not, check for good electrical connection from the knob through the bolt through the paper clip to the aluminum foil leaves. It is also possible that all the coating was not removed from the knob or that the paper clip is coated. If the leaves do not move, check to be sure they are not stuck or crimped together.

Use the electroscope until it is mastered. The next chapter deals with generating a static charge in considerably more detail. After reading through it, the reader should be able to use the electroscope efficiently, charging it with several different materials.

# Chapter 2

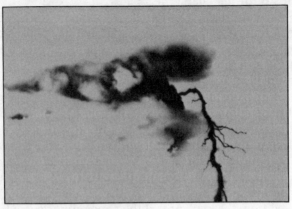

# Piezoelectricity

**P**IEZOELECTRICITY IS THE ABILITY OF A MATERIAL TO GENERATE electricity when pressure is applied or released. In other words, a piezoelectric material will generate an electrical charge when squeezed (Fig. 2-1). This is one of nature's methods of changing one form of energy into another form— mechanical energy is converted into electrical energy. The opposite effect also occurs: when voltage is applied to the material, pressure will be generated, and the size will change if the material is free to move. Both characteristics are used commercially. The ability to change size with the application of a voltage is used for such things as piezoelectric speakers, ultrasonic cleaners, and depth finders (Fig. 2-2). The ability to generate electricity with the application of pressure is used in piezoelectric spark generators, microphones, flameless pilot lights, and phonograph pickups.

Pierre and Jacques Curie discovered the piezoelectric effect in various substances back in 1880. They experimented with Rochelle salt, quartz, and tourmaline. These materials were used for many years. Then, as interest in piezoelectricity increased, several other materials were found to exhibit piezoelectric properties. These substances included barium titanate, lead ziconium titanate, ammonium dihydrogen phosphate, potassium dihydrogen phosphate, lithium sulphate, and dipotassium dihydrogen tartate. The first three materials

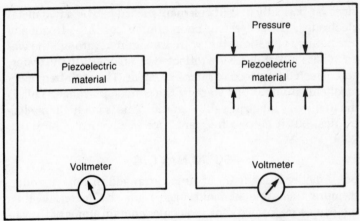

*Fig. 2-1. Demonstrating generation of voltage by piezoelectric action.*

were the first discovered and used. The titanites are commonly used at the present time. Quartz and tourmaline are naturally occurring substances, while Rochelle salt is crystallized from a solution of potassium sodium tartate. Rochelle salt, while easy to produce in crystals, is not used to a great extent. It tends to absorb moisture, is sensitive to heat, and disintegrates too easily. In spite of this Rochelle salt is sometimes used as microphones and phonograph pickups.

Piezoelectricity remained just a laboratory curiosity until about 1916 when the piezoelectric effect was used in the development of an ultrasonic submarine detector. This de-

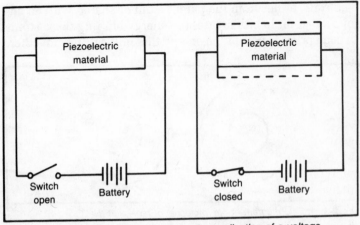

*Fig. 2-2. Demonstrating changing of size by application of a voltage.*

vice was basically a quartz element placed between two metal electrodes. It generated short bursts of high-frequency mechanical vibrations that were transmitted through the water. These vibrations were reflected by an object in the water, and detected by a second quartz element. The high-frequency mechanical vibrations were in the ultrasonic range. This is the frequency range just above audio. This is not high for electronics, but it is high frequency for mechanical vibrations.

## SOLID MATTER

Solid matter is made up of atoms arranged in some manner. If more than one element makes up the material, then molecules are formed. The molecules contain atoms of the various elements making up the material. For example, common table salt is made of sodium chloride. That is, one atom of sodium and one atom of chlorine make up one molecule of salt. The combination of the various atoms that make up the material is on the atomic level—the number of atoms of the individual materials required to make up one molecule of the material is joined on an atomic level. The molecule is the smallest division of material possible while maintaining the integrity of the material. There are millions of molecules in one grain of salt. Figure 2-3 shows the atomic diagram for one molecule of table salt.

The atoms of a solid material are in contact with each other, and interlock by sharing electrons in the outer orbit. The normal tendency is for the outer orbit to be filled with electrons, either from the parent atom or shared with an adjoining atom. In either case, the energy binding these atoms to each other is part of the energy that holds things together.

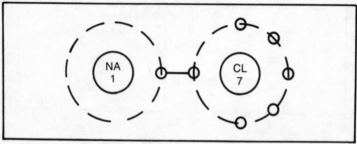

Fig. 2-3. Atomic diagram of one molecule of table salt (NaCl).

It should be pointed out that all the inner orbits of an atom have the maximum number of electrons. It is only the outer orbit that may not be filled.

There are two forms a solid material can take when it is formed: crystalline and noncrystalline. Crystalline material exists in the form of crystals. These may be small crystals visible only under a magnifying glass, or large crystals measured in feet, or any place in between. Crystals are defined as a body that is formed by solidification of a chemical element, a compound, or a mixture, and has a regularly repeating internal arrangement of its atoms, and often external plane faces. They may grow as one crystal, or as an aggregate of many either growing from a common base or growing in a geometric pattern.

A crystal has a regular, repeating internal arrangement of the atoms or molecules. They develop as solids bounded by natural plane surfaces called crystal faces. Noncrystalline materials are solids that are devoid of an orderly internal arrangement of their atoms. The same element could be crystalline or noncrystalline, depending on the conditions present when the solid was formed. For example, silicon can form large, beautiful quartz crystals, and it can also form glass. Glass is noncrystalline while quartz is crystalline, but the two materials solidified under different conditions.

Noncrystalline materials are sometimes called amorphous, which means without crystal form. For most elements or materials, the difference between crystal and noncrystalline is the conditions upon which the material solidified.

It is the crystalline materials that exhibit the piezoelectric properties. Usually the slow formation of the solid material, such as slow cooling of a molten material or slow precipitation of a material from water, is responsible for the crystals. Figure 2-4 shows a common quartz crystal. Notice the many flat faces.

At the crystal planes, ions are easily formed. This is because the energy bonding the atoms across the crystal plane is less than that bonding the other atoms. The crystal plane is a plane that forms one of the crystal faces. There are many crystalline materials, and Table 2-1 lists several, along with the chemical composition. A detailed study of crystals is very interesting, but is beyond the scope of this book.

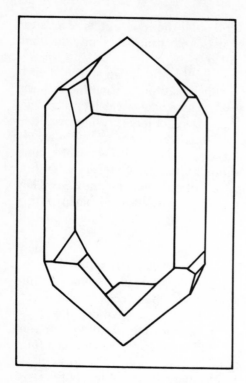

*Fig. 2-4. A quartz crystal.*

## IONS

Ions are atoms with an electron added or removed. If the atom has an added (extra) electron, it is a negative ion. A negative ion has a negative overall charge because of the added electron. If the atom has one electron removed, it is a positive ion

**Table 2-1. Table of Some Common Crystalline Materials, and Their Chemical Composition.**

| Material | Formula | Crystal System |
|----------|---------|----------------|
| Quartz | $SiO_2$ | Hexagonal |
| Zircon | $ZrSiO_4$ | Tetragonal |
| Corundum | $Al_2O_3$ | Hexagonal |
| Diamond | C | Cubic |
| Rochelle Salt | $KNaC_4H_4O_6 \cdot 4H_2O$ | Orthorombic |
| Galena | PbS | Cubic |
| Tourmaline | Complex | Hexagonal |
| Zinc Blende | ZnS | Cubic |
| Silver | Au | Cubic |

with a positive charge. This is illustrated in Fig. 2-5, which shows a positive and negative ion, and their formation. Notice that the total number of electrons does not match the positive charge number in the nucleus (the number of electrons in the outer orbit). This is because the charge on the ion is not neutral.

There are two ways an ion can move; the total atom can move, or the charge can move. In a gas or liquid, the atoms are relatively free to move, so the ion atoms can move. But in a solid, the individual atoms cannot move without breaking up the material. In a solid, the charge moves. For negative ions, the "extra" electron can move from atom to atom, if it is given the incentive. Normally this is provided by an electric field, as shown in Fig. 2-6. An electron moves from atom to atom, in the direction of the positive electric field. This is indicated by the numbered sequences in the figure.

A negative ion moves in much the same manner, except it is the absence of an electron that moves from atom to atom. This is illustrated by Fig. 2-7. Notice that for both Figs. 2-6 and 2-7, it is a charge that physically moves.

In a piezoelectric material, the ions can be moved more easily along some axis than others. Pressure in certain directions results in a displacement of ions such that opposite faces of the crystal assume opposite electrical charges. This is illustrated by Fig. 2-8, which shows the generation of ions and electrical charges by the application of force to a crystal. If

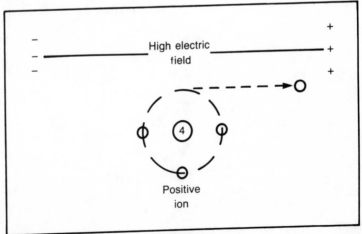

*Fig. 2-5. Ions and how they are formed.*

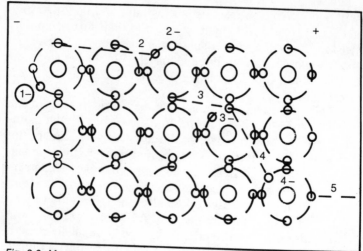

Fig. 2-6. Movement of a negative ion in a solid.

a high-impedance electrical circuit is connected to these crystal faces, a measurable current will flow. Note that when all the excess electrons at the crystal face flow to the opposite face, no additional current will flow, and the voltage will drop to zero. This happens when the force becomes steady, and no longer changing. To provide more electrons available for current flow requires the generation of additional negative ions at the crystal surface. This requires additional pressure applied to the material. So the piezoelectric voltage is gener-

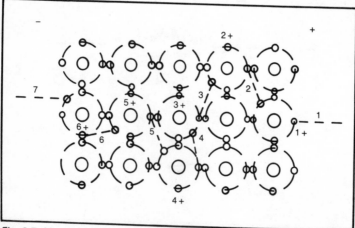

Fig. 2-7. Movement of a positive ion in a solid.

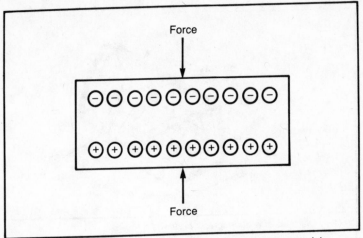

*Fig. 2-8. Generating and moving ions within a piezoelectric material.*

ated only when the applied pressure is changing.

When the pressure is released, the ions will return to their original positions within the crystal, creating an electrical charge of the opposite polarity. Again, current will flow only when the force is changing.

This is one of the theories of why the piezoelectric effect works. The exact cause is not known. Another of the theories is that, when no pressure is applied to the material, the electric charges in the crystal domains are spaced so that no net electric charge appears on either surface. That is, there are an equal number of positive and negative ions at the surface, giving a net charge of zero. Under pressure, the crystal domains are aligned so that opposite electrical charges appear on the two surfaces.

## APPLICATIONS

Pressure may be applied and released by any of several methods, and is dependent on the application or circumstances surrounding the changing of force. How great the pressure is, and the size of whatever is applying the force determines the time rate of the changing force. This, in turn, determines the frequency of the response.

If the piezoelectric material is the element of a phonograph cartridge, it is the movement of the needle that provides the changing force. The needle follows the groove on

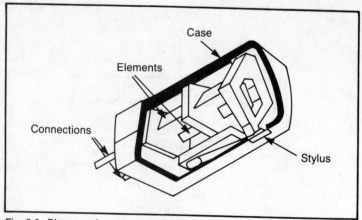

*Fig. 2-9. Diagram of a phonograph cartridge.*

the phonograph record. This groove changes depth as a function of the information recorded on the record. The needle is coupled to the piezoelectric element, as illustrated in Fig. 2-9, which is a diagram of piezoelectric phonograph cartridge. As the needle moves up and down, the force changes on the piezoelectric element. This generates a changing voltage which is amplified and becomes the sound reproduced by the speaker. So, in the case of the phonograph cartridge, it is the movement of the needle that causes a varying force to be applied to the piezoelectric element.

Another important use of piezoelectric materials is an audio tone generator. Figure 2-10 shows the cross section of such a generator. This device is usually referred to as a squealer, and is used extensively to give the "beeps" to computers and keyboards. The squealer consists of a disk (usually of beryllium copper) with a ceramic piezoelectric material bonded to one side. The ceramic piezoelectric material is manmade and not found in nature. It is used because it has good proper-

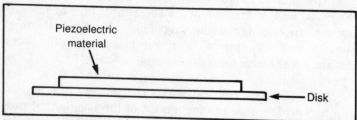

*Fig. 2-10. Cross section of an audio squealer.*

36

ties and is an easily formed powder. The disk is 1 to 2 inches in diameter, typically. The physical size of the disk is one of the limiting factors for the maximum volume. When an audio signal is applied to the disk, it will flex up and down, producing sound. This is illustrated in Fig. 2-11, which shows the squealer disk vibrating.

An audio signal is an electrical signal that changes amplitude at an audio rate. When the signal is at zero, the disk is in the flat, or neutral condition. When a voltage of one polarity is applied, the disk will flex in one direction. When the voltage is the opposite polarity, the disk will flex in the opposite direction. This movement, as shown in Fig. 2-9, moves the air molecules in contact with the squeeler at a varying rate. It is this varying rate that produces sound. The reason the disk flexes is that as voltage is applied, the radial dimension of the piezoelectric material changes. Being bonded to one side of the disk, a change in the material radial dimension will be coupled to the disk.

If the piezoelectric material radial dimension increases, this will cause the disk to bow upwards because the outside of the disk is fixed in size and not covered with piezoelectric material. Also, because the material is bonded to one side of the disk, any change in dimension to the piezoelectric material will be coupled to one side of the disk. The converse is true if the radial dimension of the piezoelectric material becomes smaller: it will try to make the mating surface smaller, bowing down the center portion of the disk.

Sound is a pressure wave travelling through air. This causes the number of air molecules of air to become compressed in front of the pressure wave, with rarified behind it, as illustrated in Fig. 2-12. The single line of air molecules is representative of the density of the air molecules at any given point. The waveshape at the bottom gives the applied

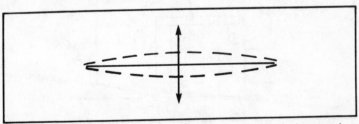

*Fig. 2-11. Diagram illustrating how a squealer vibrates, producing sound.*

37

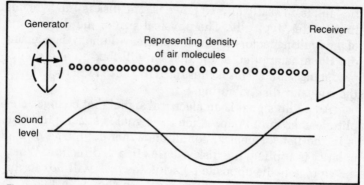

Fig. 2-12. Diagram showing the mechanics of sound generation and propagation.

audio signal. It is this changing air density that is detected as sound by an instrument (microphone) or the ear. If the density, or molecules per unit volume, does not change, no sound will be detected.

Sound is the change in the density of air molecules. This change in density results in a change in air pressure. This change in air pressure, impacting on a receiving device causes physical movement. This physical movement is interpreted as sound.

Another use of the piezoelectric affect is in ignition devices. These devices replace pilot lights in stoves and furnaces (Fig. 2-13). This contains a stack of piezoelectric materials. When pressure is applied and released by a solenoid device, an extremely high voltage is generated. This voltage causes a spark across the spark gap, and if there is a

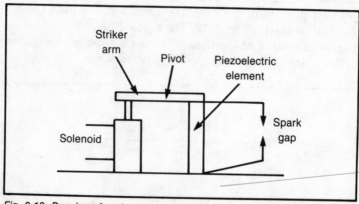

Fig. 2-13. Drawing of a piezoelectric ignition device.

combustible substance available, such as gas, a flame results. With the increasing cost of gas, and the emphasis on reducing the energy used, these ignitors are finding more and more applications. This type of application comes close to the method nature uses to generate electricity using the piezoelectric principle.

There are many more applications for piezoelectrics, which are not discussed in detail here. These include quartz frequency crystals, ultrasonic cleaning, sonar, ultrasonic gauging, and flaw detection, among the many applications. Those given here are for examples and are not meant to include all applications.

## CHARACTERISTICS

It must be noted that, if pressure is applied in one direction to a piezoelectric crystal, voltage is generated at more than one pair of opposite faces. This is illustrated in Fig. 2-14, which shows generating a charge at the three sets of opposite faces for a cubic crystal. Also note that if the crystal is placed in a shear, a charge is generated. Shear is where a par-

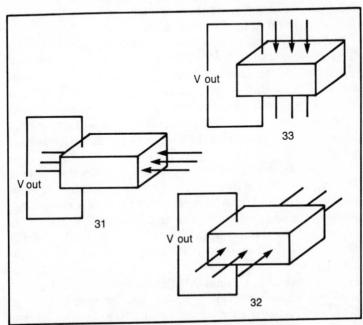

*Fig. 2-14. Generation of charge on opposite faces of a crystal.*

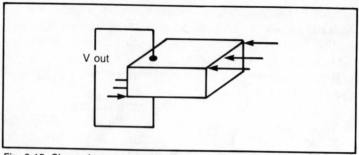

*Fig. 2-15. Shear also generates an electrical charge.*

allel force is applied to opposite edges, as illustrated in Fig. 2-15.

Tension and compression refer to the application of force. These are the most common and easily understood directions. Compression is forces applied on opposite sides, acting against each other. Squeezing something in the hands is compression. Tension is forces trying to pull something apart. These two forces are "straight line" forces, where they are applied on opposite sides and balance each other out. The net resultant force is zero. Figure 2-14 illustrates compression. If the force vector arrows were reversed, pulling instead of pushing, it would be a tension force.

With shear, the force is applied in a manner in which two parts of the material try to slide apart. The net result is a "layering" of the sample when it fails due to a shear force. Figure 2-15 shows a shear force. Note that the force is applied at opposite edges and not on opposite sides. A good example of shear force is removing old paint from a surface using a putty knife. Another example is opening the lid on a box with a sliding top.

In reality, a piezoelectric charge is generated whenever the crystal is placed in any type of stress, whether it be compression, tension, traverse, parallel, or simply two points opposite sides of the crystal. The amplitude of the voltage is dependent on the material, the pressure applied, and the direction of the pressure in relation to the electrical connections.

It is logical to assume that the charge generated on the faces where the force is applied, as in the first example in Fig. 2-14, will be larger than for the other two examples. This is the case, and it is borne out by some examples, constants, and equations. The constants involved are called the g constants and the d constants, and are defined for the three axes (the X, Y, and Z axes).

The d constant is the ratio of short-circuit electric charge developed per unit area of electrode to the applied mechanical stress. There are three d constants: d33, d31, and d32. The first number defines on which axis the force is applied and the second number defining on which axis the charge is measured. This is illustrated by Fig. 2-16. This figure shows the convention for the 1, 2, and 3 directions. Normally, the output connections are considered across the 3 direction. The other two axes are at right angles to the 3 axis. The first number in either the d or g constants is the axis of the output. It is to this axis that the output leads are connected. The second number applies to the direction of force. This is illustrated by Figs. 2-14 and 2-16.

Note that for simple compression or tension, the numbers 1, 2 and 3 refer to the X, Y, and Z axes respectively. For this example, the pressure is always applied in the Z direction. Although the constants exist for applying the force on other axes, we will take the realistic, although simplistic, approach that the other constants will be identical to those given for the Z axis.

For the case of applied shear, different constant numbers are used. For shear, the number 4 refers to shear around the X axis, 5 around the Y axis, and 6 around the Z axis. This

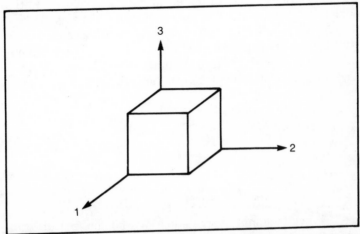

*Fig. 2-16. Diagram to define force and charge directions for the three constants.*

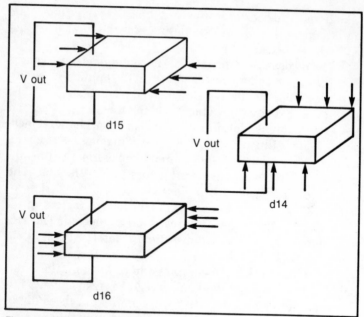

*Fig. 2-17. Diagram to define d14, d15, and d16.*

is illustrated in Fig. 2-17.

The g constant of a piezoelectric material indicates the ratio of electric field generated along a given axis to the stress applied along (or around) a given axis. The numbers, as discussed above, also apply to the g constant. For example, the g33 constant is a ratio of the electric field developed in the material's Z axis to the stress applied to its Z axis.

Notice that the d constants are the ratio of electric charge developed to the applied mechanical stress, while the g constant is the ratio of the electric field generated to the mechanical stress. So the d and g constants are related by the equation

$$Q = CV$$

where:

 Q is the charge.
 C is the capacitance
 V is the voltage

The voltage and charge equations for the first diagram in Fig. 2-14 are:

$$Q = Fd33$$
$$V/T = (Fg33)/LW$$

where:

Q is the electric charge generated
F is the force
d33 is the d constant for the 33 axis
V is the generated voltage
T is the thickness of the material
L is the length of the material
W is the width of the material
g33 is the g constant for the 33 axis

The dimensions for the g constant are: volts/meter/ newtons/square meter and is usually written more simply as: X .001 meter volts/newton. The dimensions of the d constant are: meters/volts, and are usually given times 10 to the − 12. Because the d constant and the g constant are both functions of basically the same phenomena, it makes sense that they are interrelated. That is:

$$G = d/(KeO)$$

where:

g is the g constant
d is the d constant
K is the dielectric constant for the material
eO is a constant (9 × 12 to the − 12 farad/meter)

As with the g and d constants, the K constant is a function of the axis, and it must be the same as for the other two axes. Table 2-2 gives the values for some of the piezoelectric materials.

From the constants given in Table 2-2, and the above equations, the amount of voltage generated can be calculated.

## DEMONSTRATIONS

Now that we have looked at the theory and mechanics behind the piezoelectric effect, let us proceed with some demonstrations. The easiest way to show the piezoelectric effect is with

| Material | $d_{33}$ | $g_{33}$ | $k_{33}$ |
|----------|----------|----------|----------|
| Quartz | 2.3 | 58 | 0.1 |
| Lithium sulfate | 16 | 175 | 0.35 |
| Barium titanate | 149 | 14 | 0.48 |
| PZT-4 | 285 | 26.1 | 0.64 |
| PZT-5 | 374 | 24.8 | 0.675 |
| Lead metaniobate | 85 | 42.5 | 0.42 |
| Units | $10^{-12}$m/V | $10^{-3}$V/m $N/m^2$ | — |

an oscilloscope and a test fixture, as shown in Fig. 2-18. This shows an oscilloscope connected to a phonograph cartridge. Gently moving the cartridge needle with the finger produces a trace on the scope. The more violent the movement the larger the amplitude signal, and the quicker the movement the faster the signal. This is how the phonograph cartridge works to produce sound.

An oscilloscope is a nicety when doing electrical experiments, but not everyone has one available. So other methods of visually demonstrating the results are discussed in the following pages. With an oscilloscope, there will be a problem of 60 hertz-induced voltage pickup. This 60 hertz (Hz) pickup can be reduced drastically by supporting the cartridge in a metallic fixture, and grounding the fixture to the ground terminal of the scope. Do not touch any of the wiring or the car-

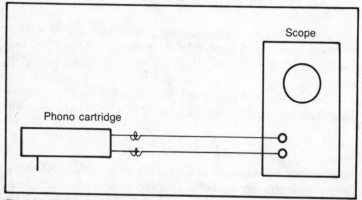

Fig. 2-18. Using an oscilloscope to monitor the output of a phonograph cartridge.

tridge. Use a wooden or plastic toothpick or rod to move the needle.

Using such a test setup, a voltage of 4 volts peak to peak, with a duration of several milliseconds, can be observed when the needle is moved. The pulse duration is controlled by the speed of the movement imparted to the needle. The amplitude is a function of the distance the needle is moved. Following are some other methods of demonstrating the presence of piezoelectric voltage.

### Headphones

It is possible to use a set of sensitive headphones connected across the cartridge, but in this case the sound level produced will be very small. It can be heard in a quiet room. Most likely it will sound like a very faint rumble. If headphones are used, the amplifier described later in this chapter will be required.

### Meter

By connecting a meter across the output of the phonograph cartridge, some indication may be obtained, depending on the type of meter used. If it is a common multimeter with the needle, the indication will be only a slight wavering of the needle when measuring voltage on the lowest range. If it is a digital meter, readings to a half volt or more may be obtained. Realize that the readings will be displayed for one sample period, then a new reading will be displayed. Depending on the meter, the sampling rate is typically a half second. Take several readings while moving the needle. In this manner the maximum reading can be obtained.

### Amplifier

If so desired and if so inclined, a simple, general-purpose amplifier can be assembled, as shown in Fig. 2-19. This amplifier will amplify the output from the piezoelectric device so that it can be heard over headphones or seen on a meter. This is an operational amplifier circuit, using high input impedances to match the impedance of the piezoelectric device. Assemble the amplifier on a small circuit board. Table 2-3 gives the parts list for the amplifier. Use a 14-pin socket for the integrated circuit, and carefully solder the connections. This amplifier is a standard type amplifier, and the parts

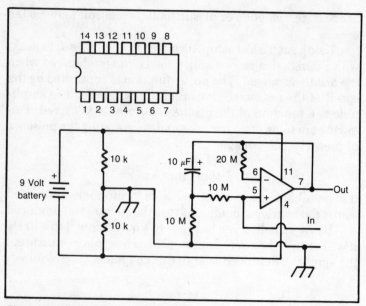

Fig. 2-19. Schematic for a simple, battery-powered amplifier for use with the phonograph cartridge and headphones.

should be available at any electronics store. The IC drawing given in Fig. 2-19, which shows the pin numbers, is a top view of the IC, as it will appear as viewed from the component side of the circuit board. Use a battery connector to connect the transistor radio 9-volt battery to the circuit. Do not connect the battery to the battery terminals until the circuit is completely assembled and checked for shorts, opens, or miswiring.

The operational amplifier used (LM324) is a quad amplifier containing four independent amplifiers. One of these am-

Table 2-3. Parts List for the Amplifier Shown in Fig. 2-19.

(1) 324 operational amplifier IC
(1) 14-pin IC socket
(2) 10K 1/4W resistors
(2) 10M 1/4W resistors
(1) 20M 1/4W resistor
(1) 10uf, 15 volt or greater, capacitor
(1) 9-volt battery
(1) Snap terminals for 9-volt battery
(1) Small circuit board

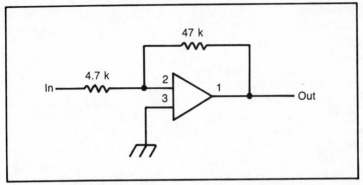

Fig. 2-20. Adding an extra amplifier to boost the overall circuit gain.

plifier sections can be used to provide additional amplification. This is shown in Fig. 2-20. This amplifier should not be required for use with the phonograph cartridge, but it might be for some of the additional demonstrations. Wire this circuit up, but leave the input disconnected until required.

To use the amplifier, simply connect it between the phonograph cartridge and the voltmeter (or headphones) as shown in Fig. 2-21. This will increase the voltage from one half volt maximum to a more readable level, with sufficient current drive to operate a regular multimeter or headphones.

These methods showing the presence of a voltage from the phonograph cartridge also give some measure or indication of the voltage amplitude. The best method is to use an oscilloscope. If one of these is not available, the next best thing

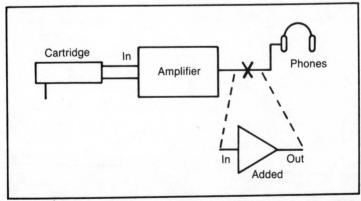

Fig. 2-21. Using the amplifier and the added amplifier.

is the digital voltmeter with a low range of 1-volt dc or less. The amplifier and a multimeter will give good indications.

## LCD

Another method of demonstrating the presence of piezoelectric voltage is using a liquid crystal display (LCD), as shown in Fig. 2-22. The LCD requires very little current to form a display, so connecting the phonograph cartridge to any two of the input pins can cause two different segments of the display to "turn on" when the cartridge needle is moved. This might appear as two black areas on a silver display.

When using the LCD to indicate the presence of piezoelectric voltages, be careful not to introduce a static charge onto the LCD, giving a false indication. Lay the LCD on an insulated surface, attach the leads using some type of clips, and make sure not to rub anything against the LCD while trying to obtain the indication.

An LCD works by applying a voltage from the input to the backplane. This voltage (3.5 volts or less) at less than a few microamps current will make a segment of the glass turn colors. This is due to the small electrical field generated between the front glass and the backplane.

The manufacturer and model number of the LCD display is not critical, so any of the many types available can be used. Make sure that it is a liquid crystal display and not a light emitting diode (LED). If in doubt, ask someone. LCD displays are available at most electronic hobby stores.

Select one of the above methods and become familiar with it because it will be used later when demonstrating the piezo-

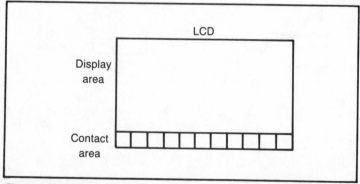

Fig. 2-22. Using an LCD display to demonstrate the piezoelectric effect.

electric effect with certain materials. The best method is to use the scope to display the generated voltage. The scope will display signals of different amplitudes and frequencies. It will also display signals of extremely low amplitude. A scope, however, is an advanced piece of test equipment, and might not be available. The next best thing is to use the amplifier with a meter or headphones. Use the "added" amplifier shown in Fig. 2-20 when required to provide extra amplification for low-level signals.

Realize that the LCD method will demonstrate the presence of a voltage. Although it does not give a good indication of the amplitude, it will suffice.

After the demonstration method has been selected and made to work with a piezoelectric cartridge, it is time for the next step: to set up a fixture to demonstrate the piezoelectric effect in various materials.

Obtain a piece of quartz. You can find the address for a mineral supply company either from the phone book or from a magazine at the library. Several cities have mineral supply or lapidary stores. If there are none locally that have quartz crystals for sale, obtain a catalog from one of the mineral or crystal supply companies listed in a mineralogist magazine. In any case, try to obtain a decent quartz crystal at least 1 inch long and well formed. Figure 2-23 is a photograph of some quartz crystals of various sizes. Notice that they are not all fully formed crystals, but they all produce piezoelectric voltage.

You could also obtain some surplus electronic crystals—preferably the type with the small screws in the front of the cans, with a frequency of 3 to 5 megahertz. Carefully dismantel the crystal holder and recover the quartz.

Figure 2-24 is a photograph of one type of test fixture. The test fixture must hold the specimen securely, and be capable of transmitting a mechanical blow to the specimen. It must also provide two isolated contacts on opposite sides of the crystal for connection to the monitoring device. The fixture shown is a clamp-on vice with Teflon-lined jaws. This type of vise is used frequently for electronics assembly work. This is only one type of holder. Any holder that meets the basic requirements will work. If a holder is used, do not clamp the specimen tight. Any pressure applied from the holder will pre-load the sample, reducing the change in pressure from

Fig. 2-23. Photograph showing some different size quartz crystals.

Fig. 2-24. Photograph of test fixture.

50

a given force. In addition, the sample can be shattered with a blow.

It is possible to tape the leads to the specimen. Use a wide conductor, such as a piece of aluminum foil, to contact the sample. Make sure that the foil is tight against the sample, and taped securely. The foil contact should be as large as possible, and against opposite sides of the sample. If this is done, there might be a problem with tapping the leads when applying the pressure to the sample.

Another possibility for small samples is to use a wooden spring clothespin. Attach a conductor to the faces of the jaws of the clothespin. In this case, tap the specimen directly to apply the force. When not using the metallic holder, work on a piece of metal that is grounded. This metal can be covered with a piece of paper if desired. This will reduce the 60-Hz pickup.

Care must be taken to prevent inductive pickup from being interpreted as the piezoelectric voltage. There were times when the author had excellent readings, only to find that it was 60-Hz pickup from the hand that was holding the tool tapping the sample. If there is any doubt, place the tool against the sample slowly, without generating any impact. There should be no reading. Also, the reading should be present only for a short duration, then go to zero. The time the reading is present is a function of the detection method used. For a digital multimeter, it will be there longer than for a scope because of the scan rate of the multimeter.

The basic frame of the vise is metallic, so it will transmit any induced shock to the specimen. The Teflon-covered jaws are lined with a conductor, which is connected to the display device, which is an oscilloscope in the author's test setup.

When applying the force to a piezoelectric specimen, it is best to tap the material directly on one of the crystal faces that is attached to the lead. Support the opposite side. If the holder is a metal vise, it is possible to tap the jaws of the vise. This impact will be transmitted to the material through the metal jaws. If the specimen is being tapped directly, use a piece of hard wood. This prevents any inductive pickup from a metallic tool or from the user. If a vise is used, a metallic tool can be used, as long as the tool is tapped against the vise frame and not against the specimen directly.

The basic demonstration is to hold the quartz sample in

the holder and to tap the sample and observe the voltage produced. Always make sure that the leads to the display device are on opposite sides of the sample, and are flat against the sample. When tapping the sample to produce the piezo-electric voltage, always tap first in the faces that the leads are attached to. This typically will produce the largest voltage. Once this is established and demonstrated, try supporting the sample on one of the other sides, and tapping on the side opposite that which is supported. This demonstrates that voltage is produced across other pairs of faces. The difference in amplitudes is readily apparent when using a scope.

The sample must be tapped with a hard object. If it is wood, it must be hard wood, such as oak. This is because the area under the response curve is constant for a given material and a given force. This is the electrical charge generated by the change in force. The maximum voltage generated is a function of the time duration of the charge. This is illustrated in Fig. 2-25. The shorter the time for the change in force, the higher the voltage.

The hardness of an object determines the time duration of an impact. The harder the object, the shorter the time. This is because there is very little, if any, deformation of the object. A soft object, such as soft wood or aluminum, deforms with the impact. This creates a longer time duration, and consequently, a smaller maximum voltage.

Figure 2-24 shows the difference in peak voltages for a relatively small difference in time durations. If there is a larger difference in times, the difference in peak voltages will be con-

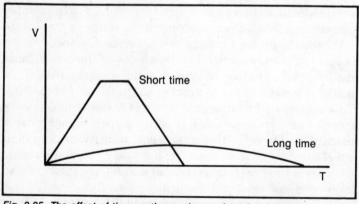

*Fig. 2-25. The effect of time on the maximum piezoelectric voltage.*

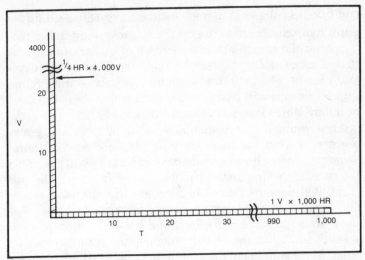

Fig. 2-26. Waveshapes for the 1,000-hour pressure application and a 15-minute release.

siderable. Suppose that the force is applied gradually over the period of 1,000 hours, and released over the period of 1/4 hour. In this case the difference theoretically is 4,000 times (Fig. 2-26). Remember that it is the change in force (or pressure) that generates the piezoelectric voltage. Earth science theory states that this is what happens during some earthquakes. The underground material is a piezoelectric material. As the pressure builds up gradually, there is no noticeable effect. But when the earthquake happens and the stress is released quickly, a significant voltage is developed. If the energy is large enough, the earthquake lights can be produced.

The basic voltage equation for piezoelectric voltage states that the voltage produced is a direct function of the length of the sample in the dimension across which the leads are attached. The larger this dimension, the greater the voltage produced. Looking at the equation for voltage generated in the 33 direction, the larger the T, or the smaller the L or W, the larger the voltage. Therefore, to produce the largest voltage, use a long, skinny crystal with flat ends. This has been proven using custom-grown crystals of Rochelle salt of different dimensions.

When growing crystals of Rochelle salt, the basic dimension ratio can be altered using copper acetate or sodium hydroxide. The copper acetate causes the crystal to grow long and thin, as compared to the "standard" crystal. Adding sodium hydroxide causes the crystal to grow short and wide.

After the piezoelectric properties of quartz are demonstrated successfully, it is time to try some other crystals. Crystals can be obtained from mineral supply stores. Some souvenir shops will have crystals, and sometimes even craft or hobby stores will have them. You could also grow a few crystals yourself. There are books in the library about growing crystals, so the methods will not be discussed here. Several different types and shapes can be grown to suit. This is a very interesting method of obtaining different crystals for additional demonstrations in piezoelectric effects.

Try as many different crystals as possible, list those that show some piezoelectric effect, and try to obtain a relative amplitude indication. In this manner, the response can be rated. Make sure to tap the sample in as many different directions as possible, keeping track of the response. Notice that there might be some crystals that show no detectable piezoelectric voltage. All of the crystals tried by the author produced some piezoelectric voltage. Granted, the response was extremely low for some, but there was an observable response. Try connecting the leads to a different pair of opposite crystal faces if you encounter any that do not show voltage.

After exhausting the supply of crystals, try some plain rocks. Pick up some common, ordinary rocks and try to determine if there is any piezoelectric response from them. Try to determine the type of rocks, and then the relative response. Determining the type of rock can be difficult; you might try to obtain assistance from the local rock club, or the high school science or geology teacher (Fig. 2-27).

The author found that several of the rocks showed a piezoelectric voltage, although most of it was extremely low. Of course, metallic samples that are electrically conductive cannot generate any piezoelectric voltage because the voltage is shorted out through the sample and never has a chance to appear at the leads to the indicating device. Sandstone, which is a matrix of sand bonded together by some material, produced good results. This is not surprising, because sand is small grains of quartz. In fact, all the quartz-based rocks tested by the author produced good results. The results were not near the amplitude of the single quartz crystals, but

Fig. 2-27. Photo of some of the rocks that demonstrated piezoelectricity.

nevertheless they were significant.

It has been proposed that the previously mentioned earthquake lights and the "spook" or "ghost" lights are caused by piezoelectric voltage. Consider an area high in quartz where the pressure is continually, but gradually, building. As pressure is applied to the rock strata, a piezoelectric charge is generated. This charge resides on the quartz and other underground rocks until it builds up high enough to form a ball of plasma. Then it partially discharges and forms a ball of electrical plasma similar to ball lightning. The energy within the ball ionizes the air, forming a colored ball that looks as if it is on fire. Thus the spook lights.

It is hoped that by the time the reader finishes this chapter, he or she is interested and motivated enough to conduct the demonstrations discussed here. The possibilities of additional study is endless. The field of piezoelectric effects on various rock formations is new and very young. There has been some discussion about using it in the field of earthquake study.

The first step in additional study is to build a good test fixture. It must be shielded so that there is no stray field pickup. Also, no stray voltages can be introduced when the pressure is applied to the specimen. Design the fixture so that a uniform force can be applied. This force can be from squeezing or from impact. Compare the results obtained from both types of pressure.

As additional work is done, many more ideas will come to mind. It is from the basement laboratory that a better understanding of piezoelectricity will come. Besides, it is fun and entertaining to experiment in an area that is not fully understood.

# Chapter 3

# Static Electricity

S TATIC ELECTRICITY IS WELL KNOWN TO MOST OF US. WE ALL
have walked across the rug and reached for the doorknob, only to have a spark jump from our hand to the knob.
We have also seen the affects of "static cling," when our
clothes cling together in the dryer. These are some of the well-
known, but hardly disastrous results of static electricity. There
have, however, been some disastrous and violent displays of
static electricity.

More than one space missile has been lost due to static
electricity, once when a large spark jumped between the
stages of a separating missile, causing the second stage (and
payload) to explode. Static electricity has also been put to
good use by mankind, most prevalently with copy machines.
It is an electrostatic charge induced on a piece of paper that
picks up dried ink for the printing process.

## HISTORY

In this chapter we will look briefly at the history of static electricity, and some of the early experimenters and their experiments, then discuss what static electricity is, and how it is
generated. Following this: methods of generating, detecting,
and measuring static electricity, plus several experiments and
demonstrations of static electricity with some set up as
projects for the reader to accomplish. The most difficult and

the most enlightning project is the Van de Graaff generator. We will assemble and use this generator.

It is up to the readers to decide what projects are to be constructed. Some projects are important to the understanding of the principles of static electricity. Other projects are of the information type, and may or may not be assembled. Before any of the projects are started, read and study the whole chapter. This will provide some insight to the principles behind the projects, and a better understanding of what the project is to do. Also read the material list and make sure all the necessary materials are available.

Electrostatics and static electricity have been known since early times. Some of the court magicians performed magic tricks for their kings and noblemen using static electricity, although it was not known what caused the effects. Up until the 17th century, a static charge was generated by rubbing two materials together by hand. In the 17th century, Otto von Guericke, of Magdeburg Germany, developed the first electrostatic generator.

This electrostatic generator consisted of a sphere of sulfur rotating on an iron axis. It was rotated by hand while rubbing against the cloth. Metal brushes were used to remove the charge from the sulfur. It was a very crude device, but allowed the user to generate a significant electrostatic field relatively easily. It also led the way for experiments in electrostatics—up until this time, static electricity was mainly used as entertainment in the courts of noblemen. Later in this chapter we will construct a small electrostatic generator.

Ben Franklin did some experimenting with static electricity. He used a hollow glass globe in his experiments. There are many other experimenters who added to the knowledge of statics. It is remarkable that, with so many contributors to the field, there are so few commercial applications. Some of the outstanding names in this field are: Varley, Van de Graaff, Volta, Francis Hauksbee, and Kelvin.

## THE LEYDEN JAR

A Leyden jar is used to store a static charge. This Leyden jar is essentially a high-voltage capacitor. The function of a capacitor is to store an electrical charge. Figure 3-1 shows the cross section of a Leyden jar. It consists of a glass jar covered

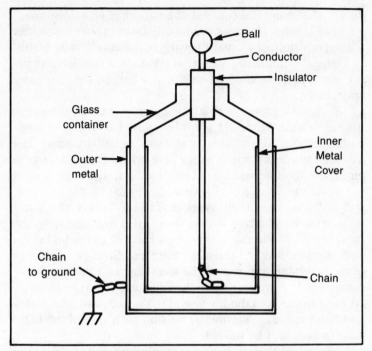

*Fig. 3-1. Cross section showing the important parts of the Leyden jar.*

with metal on the inside and outside. The mouth of the jar is sealed with an insulator, and a metal knob and rod are mounted on the insulator. A small metal chain makes contact between the metal knob and the inside metal cover. Another chain is attached to the outside conductor. This chain is used to ground the outside of the Leyden jar when it is being charged. A capacitor is made up of two plates with a dielectric material separating them. Leads are connected to the plates that allow connection to the outside world. The glass jar is the dielectric, and the metal on the inside and outside make up the plates. The knob and chain which contact the inside metal plate is one lead, and the chain attached to the outside metal cover is the second lead.

This Leyden jar was used before it was practical to generate a charge on the spot, using a static generator. A good-quality Leyden jar would actually store a static charge for a measurable amount of time. The development of static generators has reduced, if not eliminated the need of the Leyden jar. Realize that, although the Leyden jar is a capacitor, it was developed before the concept of a capacitor was developed.

The Leyden jar was used mostly by the court magicians while performing before the royalty of the day. The jar would be charged beforehand and used. While the first jar was being used, more jars would be charged behind the scenes by assistants.

Figure 3-2 gives the details about how to construct a simple Leyden jar. Table 3-1 gives the parts list. A little imagination can be used when parts are not obtainable. The insulator is a cork stopper with a hole drilled through the center. The container used is a baby food jar, although almost any wide-mouthed jar will work. Cover the inside and outside with aluminum foil. Work the foil until it is smooth and tight. Attach a chain or wire to the outside foil, and wrap the outside with electrician's tape. Punch a hole in the lid for the cork stopper. Run the threaded metal rod through the center of the insulator, and mount the metal knob on the top. Make sure to remove any coating on the knob using fine emery cloth. Attach the chain to the lower end of the rod, using two nuts and two washers. Tighten the nuts on each side of the insulator to secure it in the lid.

Make sure there is good conduction from the knob to the inner foil. This can be tested using an ohmmeter. If there is no electrical contact, the Leyden jar cannot be charged.

Static electricity is the buildup of positive or negative ions on an insulator. Static electricity is generated by the separation of two materials. They can be the same or different materials, but they are insulators. This makes sense because a conductor will not hold an electrical charge. It will dissipate the charge, usually by grounding it. Some materials tend to strip electrons from other materials and become negatively charged.

## STATIC CHARGE BUILDUP ON PLASTIC

Figure 3-3 shows how a static charge can build up on a roll of plastic material as it is being unrolled. Notice that the part being unrolled is one polarity while the roll is the other polarity. As the plastic is unrolled, the charge on the roll continues to build up. Because opposite charges attract, the roll will exert an increasing attraction for the plastic being unrolled. In this case, the charge will continue to build until something

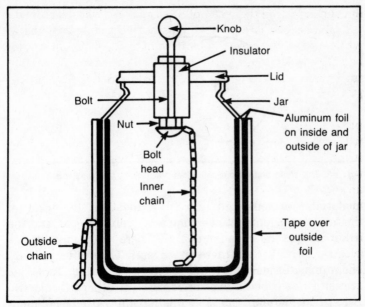

*Fig. 3-2. Constructing a Leyden jar from a baby food jar and aluminum foil.*

happens. It could be a breakdown or jamming of the equipment, or an electrical discharge. The plastic sheet that is being unrolled could exert enough force to stop the process and the sheet could wrap around the equipment.

If the material is conductive, or if a set of shorting brushes is incorporated, as shown in Fig. 3-4, the charge will not build up. This is the standard method of eliminating this problem.

## THE TRIBOELECTRIC SERIES

The triboelectric series is a list of materials, arranged from positive through zero to negative. When any two of the listed

*Table 3-1. Parts List For the Leyden Jar Project.*

Small baby food jar, glass 2 X 3 inches.
Brass knob, 3/4-inch diameter.
Bolt, with nut, 1 inch long, to fit knob.
Cork stopper, 3/4-inch diameter, cut to fit.
Conductive chain, brass, copper, or gold, length as required.
Aluminum foil, about 1 square foot.

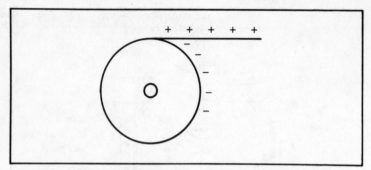

Fig. 3-3. The static charge buildup while unrolling from a plastic roll.

materials are separated or rubbed together, the material highest on the table will become positively charged, and the other accepts the negative charge. There is no "one" triboelectric series listing, there are several. They differ only by the number of materials, and the materials listed. Table 3-2 presents one triboelectric series list. The other triboelectric series lists maintain the same relationship with the common materials, such as air, nylon, polyethylene, and wool, which appear on most of the lists.

Common static charges are generated when two conductive materials separate. One material strips electrons from the other, leaving it positively charged. The hose material, with the excess electrons, becomes negatively charged. What results is known as a triboelectric charge, a charged caused by the separation of two materials or induced by friction.

The triboelectric series helps us to better understand the charge relationships of many materials. Cotton is defined as a reference material for most of the triboelectric series lists.

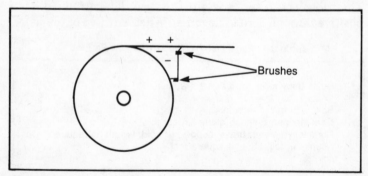

Fig. 3-4. Shorting out the static charge buildup.

**Table 3-2. The Triboelectric Series.**

| | |
|---|---:|
| Air | Positive Charge |
| Human hands | |
| Asbestos | |
| Rabbit fur | |
| Glass | |
| Mica | |
| Human hair | |
| Nylon | |
| Wool | |
| Fur | |
| Lead | |
| Silk | |
| Aluminum | |
| Paper | |
| | |
| Cotton | Zero Reference |
| | |
| Steel | |
| Wood | |
| Amber | |
| Sealing wax | |
| Hard rubber | |
| Nickel, Copper | |
| Brass, Silver | |
| Gold, Platinum | |
| Sulfur | |
| Acetate, Rayon | |
| Polyester | |
| Polystyrene foam (Styrofoam) | |
| Orlon | |
| Saran | |
| Polyurethane | |
| Polyethylene | |
| Polypropylene | |
| Vinyl (PVC) | |
| Silicon | |
| Teflon | Negative Charge |

It tends to absorb moisture, thereby rendering it somewhat conductive in terms of static charge. However, when rubbed against any other material it has the ability to induce a static charge.

The materials listed above cotton tend to take on a positive charge, or give off electrons in a friction situation. Those listed below cotton tend to acquire electrons and become negatively charged. For example, if nylon and polyethylene are rubbed together, the nylon will become positively charged and the polyethylene will become negatively charged. Realize that rubbing consists of multiple separations of discrete points on the materials that are rubbed together. So rubbing conforms to the requirement of separating the material to produce static charges.

## GENERATING A CHARGE IN AIR

The further apart two materials are, the better they are at generating a static charge. Even two materials on the same side as the cotton reference will generate a charge when separated. For example, aluminum and air will generate a static charge, with aluminum ending up with the negative charge and the air the corresponding positive charge. But because aluminum is a conductor, it must be isolated. That is, it must not be in contact with ground, or any conducting infinite source or sink of electrons.

This concept, that the earth is an infinite source or sink of electrons, must be understood. The earth is considered ground, and the zero reference against which measurements are made. The enormous size of the earth, as compared to the static charge generating materials used, mean that it is not conceivable to change the overall charge on the earth. Following this concept, any material isolated from ground can obtain a charge. If the material is not conductive, it can be in contact with ground, and still be isolated where the charge exists. If the material is conductive and touching ground, any charge will be conducted to ground.

Objects that are extremely large with respect to the object containing the charge can be considered an infinite ground source. For example, touching the knob of the electroscope with the finger may bleed off any charge, providing there is no charge buildup on the person.

Back to the aluminum and air example given above. The air contains water vapor, as measured by the humidity. Water vapor is conductive, so air with any humidity will bleed off the charge on the aluminum. The higher the humidity, the more conductive the air is and the faster the charge will be bled off to ground. Consider an aluminum airplane travelling in the atmosphere where the air is dense enough to generate a static charge, and cold enough to have a relatively low relative humidity. As the plane moves through the air, a static charge is generated faster than it is dissipated. This results in one polarity of charge on the airplane, and the opposite polarity in the air surrounding the aircraft. The charges in the air around the aircraft are attracted to the opposite polarity charge built up on the airplane. This helps to hold the charge near the aircraft. If the plane is not traveling extremely fast, the charge tends to accumulate near the tail section of the plane. Although some of the charge is swept away by the motion through the atmosphere, if conditions are right the charge continues to build. When the charge contains enough energy, it forms a plasma ball, similar to ball lightning. This ball is the color of the ionized air inside the ball, typically red or orange. This ball tends to follow the aircraft because they are still of the opposite polarities. The force of attraction is a measurable one for balls of 8 inches or larger.

The result of this scenario is the formation of colored balls of light in the atmosphere, very similar to some of the "Foo Fighters" that were observed throughout the world during World War II. This is not to say that this is the reason for all the "Foo Fighters," it is my theory to explain some of them. I have talked with pilots who have observed this phenomena over the Pacific. This does not take into account the "intelligent" ones encountered over Europe.

From the triboelectric series list of Table 3-2, the air will have the positive charge and the aluminum will be negative charge. Remember that the most positive material will produce the positive charge, and the most negative will produce the negative charge. This happens even if both materials are in the positive portion or in the negative portion.

## VOLTAGES ENCOUNTERED IN STATIC ELECTRICITY

High voltages can be developed by static charging. Presently there is no simple method of calculating the electrostatic voltage generated by an operation. There are many variables, in addition to the position in the triboelectric series. These include humidity, pressures, and speed of movement of one material over another.

Table 3-3 lists the voltage generated by some normal, everyday activities. They are listed for two different relative humidities, with the higher humidity resulting in the lower voltage. This makes sense in that the higher the humidity, the more water molecules in the air. This results in a lower

**Table 3-3. Typical Electrostatic Voltages for Low and High Humidity.**

| Means of Static Generation | Voltage | |
|---|---|---|
| | 10-20% RH | 65-90% RH |
| Walking across carpet | 35,000 | 1,500 |
| Walking on vinyl floor | 12,000 | 250 |
| Worker at bench | 6,000 | 100 |
| Clear vinyl envelope | 7,000 | 600 |
| Common poly bag | 20,000 | 1,200 |
| Work chair padded with foam | 18,000 | 1,500 |

resistance and the charge bleeds off faster. The voltages are measured with static charge meters, or electrostatic voltmeters. These meters are designed to require no current for operation, and usually measure an affect of the static charge.

## ELECTROSTATIC VOLTMETER

Figure 3-5 shows the details of a simple, electrostatic voltmeter. It works on the principal that like charges repel. The metal vanes with needle is the moving assembly. It pivots on a rubber band stretched around the pivot fixture. In the full-scale position, the vanes are just above, but not in contact with the two metal plates. The input is connected to the metal knobs. Usually, one of the inputs is grounded and the other

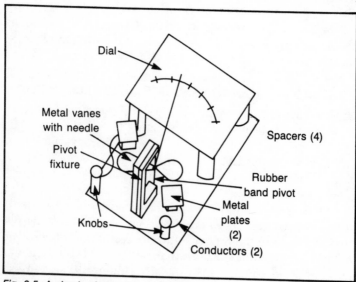

*Fig. 3-5. A simple electrostatic voltmeter.*

knob used as the input. The knobs are connected to the metal plates by the conductors.

The rotating vanes and the fixed metal plates are all charged with the voltage induced at the input. If the left-hand knob is charged positive, and the right-hand knob grounded, the charge will be induced into the moving vane. The positive charges in the vane will be moved to the right side and the negative charges to the left side. This happens because the positive charge at the left-hand metal plate will repel the positive charges and attract the negative charges.

The force of attraction causes the needle to pivot on the rubber band pivot. The amount of movement of the needle is a function of the induced charge. This meter is shown for information, and as an example of using the force of repulsion between like charges. If so desired, the reader can build the instrument by expanding on Fig. 3-5. There are other, simpler devices that show the presence of a static charge. These will be used in this book.

For a spark to be felt requires a minimum of 3,500 volts. This means that if you felt the spark when you touched the doorknob, you were charged to at least 3,500 volts. Now, 3,500 volts sounds like quite a lot, and it is. But the charge, or amount of current, is extremely low. The charge is a measure of the number of electrons that are available, and that contribute to the spark. It is the charge, or electron flow, that causes electrocution when electric lines are touched. The chances of getting electrocuted from a static discharge are nil, but other damage can be done. The spark can cause fire, or ignition of an electric detonator, or electrical interference. The discharge of the high voltage can cause damage to electronic circuits.

## DETECTING THE PRESENCE OF A STATIC CHARGE

Many things will indicate the presence of a static charge. The electrostatic voltmeter mentioned earlier in this chapter and the electroscope described in Chapter 1 are two of them. In addition, there is the suspended styrofoam ball, cigarette ashes, small pieces of paper, neon bulb, liquid crystal display (LCD), and hair, among other things. The electroscope, constructed in Chapter 1, is the common indicating device. The reader should master the use of this device. Realize that if the humidity is high, the device will not work decently. Also remember that the leaves are made from aluminum foil, which oxidizes easily; aluminum oxide is not conductive and might hamper the operation. This will cause problems where the leaves are suspended from the hanger. If this area oxidizes, the charges cannot move freely down to the bottom of the leaves.

Another thing to be aware of is that there can be a charge on the operator, so if the electroscope or any other device is touched in hopes of grounding the charge, the device will obtain the same charge as the operator. This can be remedied by touching a ground occasionally. This ground can be a wire connected to a water pipe, or the metal case on some grounded appliance of electrical device.

A small neon lamp will detect the presence of an electrostatic potential of 90 volts or more. The ionization potential for a Ne2 type neon bulb is 90 volts. A dramatic demonstration of static voltage is done by placing a Ne2 neon bulb in a small clear plastic bag, as illustrated in Fig. 3-6. Make sure the leads are spread as shown in the figure. By moving and rubbing the plastic, the neon bulb can be made to fire. This must be done in a darkened room because the pulses of light are short. Obtain the neon bulb from an electronic hobbyist store, and try this demonstration.

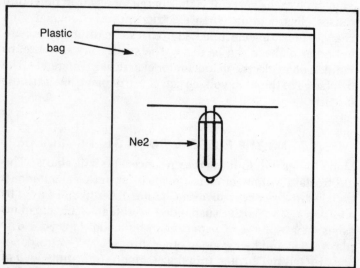

Fig. 3-6. Using a Ne2 neon bulb to indicate the charge generated in a plastic bag.

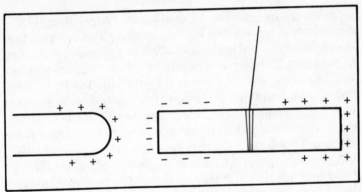

*Fig. 3-7. Using a suspended piece of styrofoam.*

The LCD can be the same one as used in the piezoelectric demonstrations. Because the LCD operates at such a low voltage, it will operate easily if the conditions are right. Lay the device on a flat surface and hold the charge a short distance above the front. If there is a charge difference between the flat surface and the charged object, some of the segments will become visible. It might be necessary to use one of the objects that were rubbed together to produce the charge as the surface to lay the LCD.

A suspended piece of styrofoam makes an interesting indicator of static charges. Suspend a small piece, about 2 inches long and about 3/4 inches wide from a thread suspended so that it is free to move in all directions. Figure 3-7 shows such a device, with a positive charge placed near one end. The positive charge forces the positive charges on the styrofoam to the opposite end, thus there is attraction between the external object and the styrofoam piece. Now, if the charged object is quickly moved to the opposite end, there will be a repulsion, and the styrofoam piece will pivot on the thread. If the charges are strong enough, it can pivot 180 degrees. This is back to the original position, so they will attract each other.

This is the modern-day replacement for the pith ball. The pith ball was light and suspended from a frame and moved anytime a charge was brought near. This styrofoam piece can be suspended from a frame made from a coat hanger. This setup is portable and does not require thumbstacks in the ceiling.

The styrofoam can be charged by grounding one end of the piece while the charge is close, as shown in Fig. 3-8. This conducts the excess positive charges to ground. Remove the ground, then remove the external charge. The result is that the styrofoam piece is charged. This charge can be checked using small bits of paper or any of the indicating devices.

One of the old standbys is the pith ball. This is a small, lightweight ball that the demonstrator would make respond with considerable movement. Pith is the soft inside of a plant stem. To make a pith ball, cut a plant stem and obtain some of the soft inner part. Carefully cut this in the shape of a ball, and dry thoroughly. Coat the pith balls with aluminum or gold paint and attach a silk thread about 6 inches long. Make a stand from wood or wire to support the pith ball and allow it to swing freely. Bring objects rubbed with silk, fur, or flannel near the pith ball and watch how it behaves. Notice that it is first attracted then repelled when the pith ball is allowed to touch the charged object.

An interesting device is created when two pith balls are suspended from the same frame. When both pith balls are charged the same polarity, they will repel each other. The distance, or angle of repulsion is a rough measure of the magnitude of the electrostatic charge. Try experimenting with this device, charging it in various ways, and making it behave as an electroscope.

Everyone has experienced, at one time or another, the sensation of hair being attracted to the comb, and following it. This happens when the hair is extremely dry, and can be cured with a little water. But this is static attraction, and works in a principle very similar to that discussed above. A good, easy indicator is to move the charged object back and forth

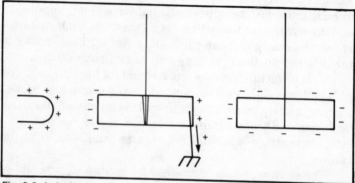

Fig. 3-8. Inducing a charge on the suspended styrofoam.

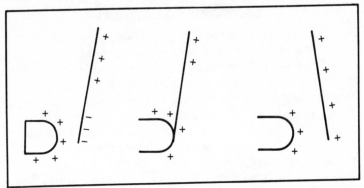

Fig. 3-9. Introducing a charge on a thread by contact.

about 1/4 inch above the hair on the arm. You can feel any hair movement, which indicates a difference in electrostatic charge.

Almost everyone has seen the demonstration where in small bits of paper are picked up with a charged object. The pieces of paper must be small enough to be picked up. Larger pieces require larger charges to lift the total weight. Also, the paper must be dry. Even a little moisture will bleed off the charge rapidly.

It takes a static charge of about 4,000 volts to pick up cigarette ashes 1 inch long, so holding a charged object 1 inch above ashes, will attract the ashes if the charge is at least 4,000 volts. This gives a quick indication if the charge is in excess of 4,000 volts.

Another simple indicator is to suspend a piece of thread. This is lightweight and easily moved by a charge. Referring to the triboelectric series given in Table 3-2, it can be seen that almost any thread will work: polyester, cotton, nylon, or silk. Because the thread is light, it will move rather large distances with relative ease. The force can be attraction or repulsion, depending on the initial conditions. If the initial movement was attraction and the thread and the charged body touched, the force could be changed to repulsion. This is illustrated in Fig. 3-9. Two things to remember when analyzing this figure: First, the charge is generated on the object and it is moved near the thread, so initially the thread has no charge. Second, there is more positive charge on the object than negative charge on the thread, so all the positive charge is not neutralized when the two items contact.

## STATIC DEMONSTRATIONS

There are several static electricity demonstrations that can be done. Most of us have seen demonstrations while in school; some impressive, some more mundane. Simple demonstrations are presented here. They are meant as an aid to understanding static electricity, and not as parlor tricks. It is recommended that the reader perform most of them.

Cut two strips of newspaper, about 1 inch wide and 8 inches long. Hold them between the thumb and fingers on one hand, as shown in Fig. 3-10. Stroke them lengthwise, from top to bottom, with the thumb and finger of the other hand. The bottom of the papers should spread. From the triboelectric series it can be determined that the paper will have a negative charge with respect to the hands. The paper will be held apart at the bottom by the negative charges. This is very similar to how an electroscope works.

Holding one of the pieces of paper, bring the free end close to the front of the TV picture tube while the TV is on. The paper will be attracted to the face of the tube, as in Fig. 3-11. The picture on a picture tube is formed by electrons striking the phosphor coating on the inside face of the tube.

*Fig. 3-10. Using two strips of newspaper to demonstrate static electricity.*

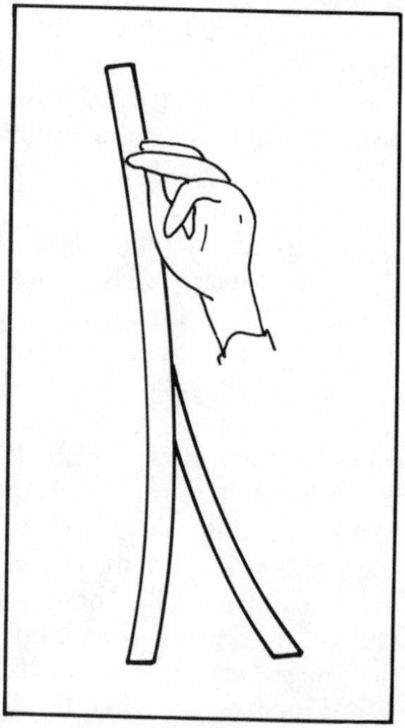

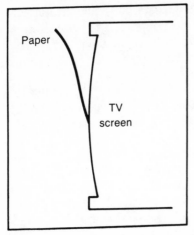

Fig. 3-11. *The attraction between a strip of newspaper and a TV screen.*

This results in an excess of electrons on the tube, which translates to negative charges. The paper is attracted because the negative charges on the paper are repelled, leaving positive charges in contact with the tube face. The force will be strong enough to hold small pieces of paper against the face, with no other support.

Most TV screens have a layer of dust on them. That is, unless the room is kept extremely clean and dust-free. This layer of dust is attracted to the screen by the static charge generated of the picture tube face. Electrostatic precipitators work in the same manner; static charges attract the dust and smoke particles and remove them from the air.

Move around the house, using the single strip of newspaper to check for a static charge buildup on various items. Try glass windows, wool blankets on the bed, and anything that comes to mind. Some might attract the paper while others do not.

Obtain a fluorescent light bulb. Rub this bulb with a piece of fur or flannel in a darkened room. Notice that whisps of light appear as the gas inside tries to ionize. Do not squeeze the bulb hard enough to break it.

Rubbing a balloon with fur, or against hair or a wool sweater, and placing it on a wall, will cause it to stick to the wall in one position.

## GENERATING A STATIC CHARGE

As previously discussed, a static charge is generated when

two non-conductive materials separate. This may involve rubbing of a hard object with a soft object, or the unrolling of a plastic. Realize that the act of rubbing is moving of two objects over each other. When two objects move, various points on each separate with respect to each other.

A hard and a soft object are usually used because the surfaces must be in intimate contact. If both objects are hard, this is difficult. Only with a soft object can intimate contact be achieved.

Several methods of generating a static charge have been alluded to. A convenient and easy means of generating a good static charge is using "magic mending" or "magic transparent" tape found in the stationary or school supplies department at the local store. This is the tape that appears opaque white until applied to paper, then it becomes invisible. A 3/4-inch-wide roll, about 2 1/2 inches in diameter, is a convenient size to work with. The smaller rolls can be used, but they are harder to work with. Figure 3-12 illustrates using the tape to pick up some small bits of paper.

To use the tape, quickly unroll about 7 inches. Use the back, or non-sticky, side as the charged side. Although both sides are charged, using the sticky side can be messy—especially when something sticks to it. The piece of tape can be reused as long as, when the tape is rewound, it sticks flat. When the stickiness of the tape is gone, or the tape is crumbled, tear off the piece and start with a new piece. In this manner, one roll of tape can last for hundreds of times.

To get a feel for how long this takes, unroll a piece of tape and hold it until the charge dissipates. Use small pieces of paper to determine when the charge has dissipated: The point at which the tape will hang limp, and not try to roll

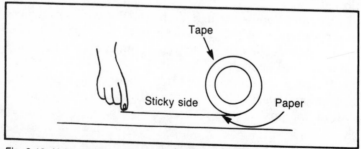

Fig. 3-12. Using "magic mending" tape to generate a static charge and pick up small bits of paper.

up. When this time has been determined, unroll a piece of tape and quickly rub it on a piece of wool. Note that this dissipates the charge considerably faster than just holding it does.

The unrolling of the tape creates a situation as depicted in Fig. 3-3, the unrolling of a roll of plastic. The unrolled tape and the roll of tape acquire opposite charges, with the dividing line at the point of separation. The separation must be fast for the charges to be separated. The adhesive on the tape facilitates this, in that it makes the separation of the tape from the roll a single, quick motion. To illustrate this point, unroll a section of tape at different speeds, from extremely slow to the fastest possible. Note that the intensity of the charge varies from zero for extremely slow, to relatively high.

Practice with the tape, unrolling it as quickly as possible for the short section. It seems that the faster it is unrolled, the better the charge. Practice—picking up small bits of paper—until the process is mastered. Notice that the paper will usually stand on one end before being fully attracted to the tape. This is because the opposite affect is happening at the opposite end of the paper bits. The lower end of the paper and the surface on which the paper is resting are of the opposite charge, and therefore attracted. If the surface is a charged, non-conductive surface, such as a wool blanket, a charge may exist between the paper bits and the blanket. This charge may be larger than that induced by the tape, so the paper will not move. It is attracted more strongly to the blanket.

Try the tape with the suspended styrofoam piece, and with the suspended thread, both discussed earlier. Extremely good results should be obtained. For the styrofoam piece, let the styrofoam contact the tape and observe the results. Notice, when using the tape, that it has a tendency to roll up if the end is released. This is because of the difference in charge on the unrolled portion and on the roll.

Try picking up dust and lint with the static charge on the tape. Unroll it quickly and place it close to the dust. The dust will be attracted to the slick side of the tape by the static charge. If there is a smoker in the family, try it with cigarette ashes. To attract cigarette ashes from 1 inch above requires about 4,000 volts static charge. Practice until this is mastered, but make sure the ashes are on a conductive medium so that no charge builds up on the holder.

Notice that when a length of tape is quickly unrolled and the end released, the end will reroll onto the tape roll. This is due to the static attraction between the end of the tape and the roll. When this has been successfully demonstrated, hold the tape and roll so that the end of the tape must rise against gravity to reroll. The tape will do this when it is charged. Try unrolling some tape and quickly rubbing it with a wool blanket. This will discharge the tape, causing it to hang limp instead of rolling up on the tape. To reintroduce the charge, roll the tape up and gently rub it with the fingers. This will redistribute the charge, so that the tape will stick to the roll by the electrostatic energy.

When the technique has been mastered, and the reader has made the electroscope respond at will, try the following demonstration to illustrate the affect of humidity. Remove the electroscope from the jar and support the lid so that it can be used. Next, using the tape, make the electroscope leaves separate because of the charge. Using a fine mist sprayer, spray some water in the area of the leaves, and try again. The response will be considerably less, or none at all. This is because the water droplets absorb the charge, and bleed it off.

A plastic pill bottle, when rubbed with a piece of fur, leather, or felt, will produce a charge. Try experimenting with several different types of small plastic bottles (the orange ones 1 1/2 inches in diameter by 4 inches long worked best). They are always available from the local pharmacy, if there are none on hand. To use, wrap the material around the bottle and rub back and forth, maintaining medium pressure. This charges the sides of the bottle. To charge the bottom, rub the bottle over the material in a quick back and forth motion. It is not required to charge both the sides and bottom.

A piece of styrofoam is a good candidate for static charging. Make sure the piece is dry, and rub it with a piece of leather, felt, or silk. In fact, it can pick up a charge from almost any non-conductor; even rubbing it with a hand will work, as long as everything is dry.

A plastic rod rubbed with a woolen cloth will produce a negative charge on the rod. A glass rod rubbed with silk, fur, or rayon will produce a positive charge on the glass rod.

Many other things can be used as static generators: a rug, a wool sweater, a hard rubber comb—and the list goes on and on.

The following is a method to generate sparks from static electricity. Obtain an aluminum cake pan about 8 inches square. Heat the metal evenly over a flame. Touch a stick of sealing wax or wax candle to the center of the aluminum pan. The wax will melt and stick firmly to form a handle. Place a piece of sheet rubber, such as an automobile tube, on a table. Stroke the surface of the rubber briskly with a piece of fur or flannel for about a half minute. Place the aluminum pan on the rubber and press down hard with the fingers. Remove the fingers and lift the aluminum by the handle. Bring a finger near the metal and you should get a spark.

## STATIC GENERATORS

Static generators are a class of simple machines that generate a static charge. The charge is continuous and can be used in various experiments. The basics of the static generator is that it has a handle that is turned, rotating a main body. As this main body rotates, it rubs against a piece of leather, generating a charge. This charge is picked up by an electrical brush, and made available for use. Figure 3-13 shows one such gener-

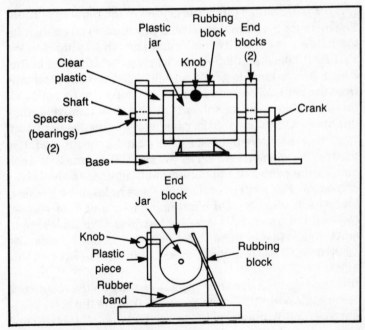

*Fig. 3-13. A simple electrostatic generator.*

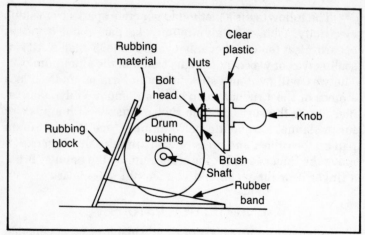

*Fig. 3-14. Some details of the electrostatic generator.*

ator. The side view is with one end block and the crank removed to show more detail. Although it is a small generator,
it will produce a continuous charge, available at the output
(knob). Figure 3-14 gives some of the details of the brush and
the leather pad.

A small, orange pill bottle is used as the rotating portion.
The distance between the two end pieces must be enough for
the bottle to turn easily, and room enough to glue it to the
shaft. All labels and tape must be removed from the bottle.
Use an X-acto knife to gently peel the label and any clear tape
from the bottle. The residual adhesive must also be removed.
Be careful about using solvents or chemical removers, they
might react with the plastic, causing it to dissolve or distort.

From these two figures, a device can be constructed. Use
a 3/16-inch diameter dowel as the shaft. Drill one 3/16-inch
hole in the center of the bottom, and another in the center
of the cap. The shaft goes through these holes during assembly. Attach some type of offset crank to one end, as shown.
Use two hollow metal spacers as bearings. The shaft should
move freely inside them, but there should not be sidewise
movement. These bearings are secured inside the wooden side
blocks.

Here again, make sure that any coating on the brass knob
is removed with fine emery paper. Make sure the bolt is conductive, and is the proper size for the knob. A rubber band
is used to hold the block with the rubbing material against

the plastic jar. This block is not secured at the bottom, and rests against the bottom block and the rotating jar. The rubber band goes under the jar, and loops around the front of the bottom block. A piece of clear plastic is used to mount the knob. This is used because it is a better insulator than wood, thus reducing charge leakage from the brass knob.

To assemble the machine, mount the bottom piece on the base and secure the two side blocks. When mounting the side blocks, insert the shaft and make sure it turns freely. This makes sure that the two shaft bearings are aligned for easy rotation. Next, mount the knob on the clear plastic piece as shown, and mount this assembly to the side blocks. Then assemble the shaft and plastic jar in the machine. Decide which side the crank is to be on, and insert the shaft through the first side block. Place the plastic jar, with lid attached, between the side blocks and push the shaft through the holes. Make sure the shaft rotates freely, and glue the plastic jar to the shaft. The remaining items are the rubbing block and the brush assembly.

The brush assembly is clamped between the second nut and the head of the bolt holding the brass knob to the plastic piece. The brush should be flexible enough to contact the plastic jar, and rigid enough so as not to be deformed. If desired, a brush assembly can be constructed from small strips of beryllium copper, or some such flexible conductive material. As a last resort, the brushes can be made from some bare electrical wire, wound around the bolt head. The wire should be twisted to give some rigidity. At the end, the individual strands should be separated so that each makes contact at a different spot on the plastic container.

The rubbing block is a wooden piece with the rubbing material glued to the contact area. Determine the contact area, then glue the material at least a half inch on each side. Determine where the rubber band is to be placed, and cut a few small notches. This helps to keep the rubber band in place.

To use the generator, turn the crank so that the top of the plastic jar moves toward the front, as shown in Fig. 3-15. It is advisable to clamp the generator to prevent it from moving when in use. In use, the crank is turned rapidly, and the charge is removed from the brass knob. Holding an electroscope in contact with the brass knob will transfer the charge to the electroscope, causing the leaves to separate.

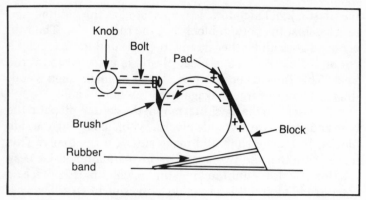

Fig. 3-15. How the charge is generated and picked up off the rotating plastic container.

Although this generator uses a plastic container to generate the charge, the rotating piece can be made of several different materials: amber, sulfur, styrofoam, wax, etc. Anything that will generate a charge when rubbed against the material mounted on the rubbing block can be used, but pill bottles were found to be the most convenient and easiest to work with, and will not come apart.

There are other types of electrostatic generators, with the above described being the simplest. The best known of the generators is the Van de Graaff generator; its construction is covered in the last part of this chapter. Another of the better known generators is the Varley electrostatic generator.

The Varley generator is not easy to construct and adjust. This machine consists of twin rotating plastic disks with aluminum foil patterns cemented to both sides. It is rotated by a small motor, and the generated charge is enough to create sparks across a spark gap.

## THE ELECTROPHOROUS

The electrophorous is a device to generate and transfer static charges. It is interesting and challenging to work with because so many things can be done with it. For example, it can be used to demonstrate the power of attraction between oppositely charged objects, it can be used to transport a positive charge on a metal disk, and it can, with patience, "float" one piece of thin plastic above a metal disk.

The electrophorous consists of two parts: (Fig. 3-16) a

rectangular block of insulating material such as Lucite or polyethylene, and a small metal disk with a plastic handle. The metal disk can be a tin can lid, about 3 1/2 inches in diameter. The handle is a plastic rod about 1 inch in diameter and 4 inches long. It is secured firmly to the metal disk. The handle is used to pick up the metal disk. It must be long enough so that the handle can be gripped without touching the disk. It must be large enough in diameter so that it can be held comfortably.

When the plastic piece is wiped with a woolen cloth, electrons are removed from the cloth and deposited on the plastic. The cloth becomes positively charged and the plastic negatively charged. This is as per the triboelectric series.

Now, if the metal disk, with handle, is placed on top of the plastic, the negative charges in the plastic will attract the positive charges in the metal disk, as shown in top portion of Fig. 3-17. As expected, the positive charges are attracted to the side of the metal plate toward the plastic, and the negative charges are repelled to the opposite side. Next, with a ground wire, the electrons (negative charges) on the plate are grounded, as shown in the middle part of Fig. 3-17. Note that this ground only touches the topside of the metal disk, as shown. Do not touch the side in contact with the plastic.

Next, the ground line is removed. The metal plate is now charged positive and the plastic negative. If the charge is great

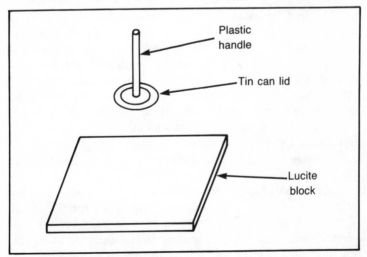

*Fig. 3-16. A simple electrophorous.*

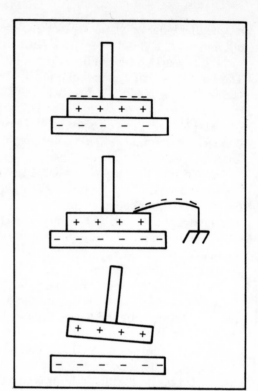

Fig. 3-17. Charging the electrophorous.

enough, the plastic piece can be picked up by the metal disk. To use the electrophorous, separate the metal disk and the plastic block by using the handle, and pulling it away at an angle. The metal plate is not a separate positive charge. Use the electroscope and the suspended styrofoam to show the presence of the charge. Recharge the electrophorous as required. When used in this manner, the electrophorous and the Leyden jar have similar applications. Also, the electrophorous can be used to charge the Leyden jar.

If grounding one side of the metal disk does not work, try touching the metal disk with a finger. This should accomplish the same end results, unless a charge has been picked up by the user.

After charging the charge carrier, or metal disk, touch it to one lead of a Ne2 neon lamp, with the other lead grounded. The Ne2 will momentarily light in a darkened room. The charge carrier can be used in most of the demonstrations using a charged object. It will move the electroscope leaves and move the styrofoam piece. It should pick up small bits of paper and even dust or hairs.

Now let us try to "float" a disk, or suspend it in the air using the force generated by static charges. Remember that like charges repel, so some means must be used to achieve two charges of the same polarity.

Make a second disk, and plastic handle assembly—only make the total assembly as light as possible. Use a hollow handle, if possible. Charge the first disk with the positive charge, as detailed above. Then charge the second disk using the positive charge on the first, as shown in Fig. 3-18. Do not let the disks touch. When the disks are in close proximity, ground the handle side of the second disk. As shown, this grounds

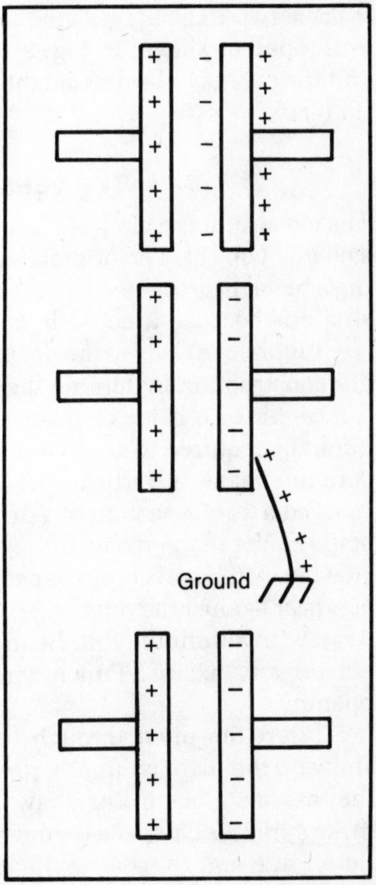

Fig. 3-18. Charging a second disk assembly with a negative charge.

Ground

83

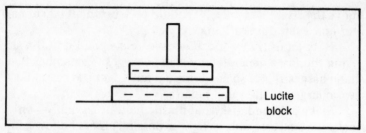

Lucite
block

the excess positive charges residing on the back side of the disk. Then remove the ground and separate the disks.

At this point, the second disk assembly is negatively charged, and the Lucite block is also negatively charged, so if these two like charges are placed in close proximity, they will repel, as shown in Fig. 3-19. Depending on the weight and the charges, the disk might actually float above the plastic block.

## THE VERSORIUM

The versorium is a simple pivoting device that moves toward a charged object. The original versorium was constructed using a metal bar, pivoted in the center. It was good for demonstrations because it could be made to turn, almost at will.

Figure 3-20 shows the details of a simple versorium. Begin construction by forcing the head of a pin or needle into the center of a balsa or other soft wood block. Wood glue might be required to secure the pin or needle in place. Make sure the shank is vertical.

Cut a 2-inch section of a drinking straw. Cut an opening at the center of the straw, through one side, leaving the other side intact. The pivot (pin or needle) will go through this hole and bear against the other side. Coat the straw with glue and wrap with aluminum foil. Bend the foil around the ends and cut off any excess. Punch through the foil at the center opening.

Insert the pivot through the opening and balance the straw on the pivot by lightly pressing the pivot point against the inside surface of the straw. Gently rotate the straw and make sure it is balanced on the point. If necessary, trim a little off one end to balance the straw.

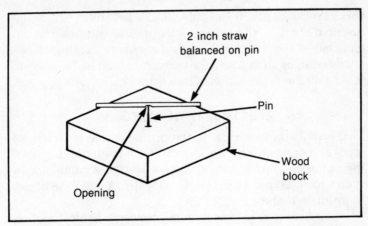

*Fig. 3-20. A simple versorium.*

The versorium works by induction, as shown in Fig. 3-21. When a charged object is placed close enough to the versorium so that the aluminum foil is within the electrostatic field of the object, the versorium rotates toward the charged object.

If the object has a positive electrostatic charge, negative charges (electrons) in the foil will move toward the charge. The attraction force between opposite charges will cause the

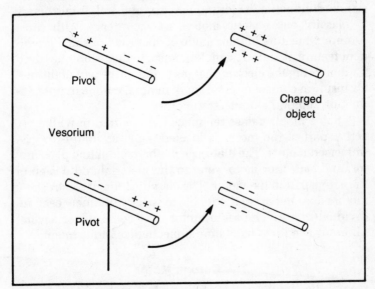

*Fig. 3-21. How a versorium works.*

versorium to rotate. If the object has a negative charge, the negative charge will move to the opposite end, leaving the end nearest the object with a positive charge. Again, the opposite charges attract, and the versorium rotates. This action is very similar to the action of the suspended styrofoam piece.

## ELECTROSTATIC MOTORS

Electrostatic motors work on the principle that like charges repel and opposite charges attract. It is a matter of introducing opposite charges onto a rotating charge accumulator in such a manner that it reacts with or against some fixed object to produce torque.

Conventional electric motors create mechanical motion as a result of magnetic forces acting upon electric currents. These motors are properly called electromagnetic motors. Electrostatic motors create mechanical motion as a result of electrostatic forces acting between electric charges.

The first electric motor invented was an electrostatic motor. Electrostatic motors have been operated from voltages in excess of 100 kilovolts. They also have been operated using currents less than .001 microamp. Electrostatic motors have been operated directly from the atmospheric electricity. The standard induction motor can do none of this. It is customary to classify electrostatic motors in accordance with some prominent feature of their mode of operation or some prominent feature of their design. Reference to the techniques used for delivering the electric charges to the active part of the motor result in contact motors, spark motors, corona motors, induction motors, and electret motors

Some motor names reference the medium in which the active part of the motor is located, such as liquid- or gas-immersed motors. The dielectric motor, conducting plate motor, and capacitor motor refer to the material and design of the active part of the motor. The classification of electrostatic motors does not make it possible to specify uniquely each individual motor. The same motor may come under several different headings by a simple mechanical adjustment.

### The Corona Motor

One of the simpler electrostatic motors is the corona motor as shown in Fig. 3-22. This is an early corona electrostatic

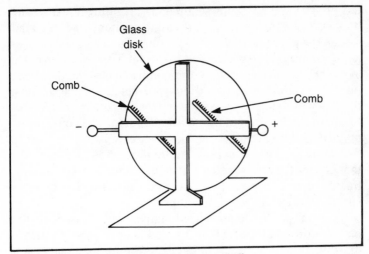

*Fig. 3-22. A corona motor as built by Poggendorff.*

motor constructed by J. C. Poggendorff to study the various parameters associated with the corona motor. It consists of a glass-disk rotor placed between two ebonite crosses carrying sharp needlelike "combs." One of the basics of electrostatics is that a sharp point works better than a smooth radius for attracting or transferring a charge. That is because the charge density (charge per unit area) is higher at the point. Remember that lightning rods are pointed. The comb consists of a metallic rod with several sharp-pointed needles attached. The sharp points almost touched the surface of the disk, so that the disk was charged by means of a corona discharge from the combs when connected as shown. Each comb could be oriented so as to make an angle with respect to the radius of the rotor. The motor will rotate equally well in either direction, and is not self-starting. However, if the combs are slanted with respect to the radii, the motor becomes self-starting and unidirectional. Combs can be mounted on both sides of the disk, with separate contacts brought out. These will be opposite the existing combs, and charged to the opposite polarity.

A corona motor, as in Fig. 3-22, could be constructed with a 45 rpm record as the rotor. Secure it to a shaft and provide some kind of bearings or bushings for the shaft. The rotor must rotate freely. The combs consist of metal rods with holes drilled through. Sharp-pointed pins or needles are used for the points, and they are secured by crimping them in the holder. The sharp needles and comb assembly are adjusted so that the points barely clear the rotor. There is a corona discharge between the points and the rotor. This transfers some of the charge on the needles and comb assembly to the rotor, immediately under the points. It is the repelling force between the combs and the charges induced by the combs onto the rotor that provides part of the rotational force. The comb of the opposite polarity attracts the charges previously sprayed onto the rotor. Each comb is repelled by the segment of the rotor that carries the charges that it sprayed onto the rotor, and attracted by the segment of the rotor carrying the charges sprayed on by the previous comb.

Corona motors come in all shapes and sizes, with the main requirement being that the charge be induced to the rotor with some sharp-pointed combs. Figure 3-23 shows another corona motor. This is a self-starting motor with a cylindrical. It is made self-starting and unidirectional by slanting the combs in the direction it is desired that the rotor rotate. This could also be used with the electrostatic motor shown in Fig. 3-22. Align the comb along a radii, and slant the comb in the direction of desired rotation. Figure 3-24 shows how the charge is induced and the rotational force

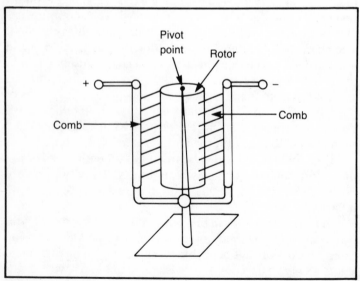

Fig. 3-23. A self-starting corona motor with a cylindrical rotor.

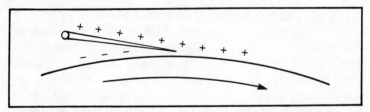

Fig. 3-24. Injecting charges onto the rotor, showing the charge relationship that creates motion.

generated in a corona motor. The charge is transferred to the rotating piece by the coronal affect. The high charge density on the point of the comb forces some of the charge off the comb and to the rotor. This charge, being the same polarity as the comb, is repelled from the comb, and a rotational motion is developed. The opposite comb is of the opposite polarity, so the portion of the rotor under the comb but not under the points, is attracted by the opposite charges.

To construct a corona motor requires a little patience, but is not difficult. Figure 3-25 shows some of the construction details, mounting the rotor to the shaft, and using bearings to mount the shaft.

The rotor is made from an old 45 rpm record with all the

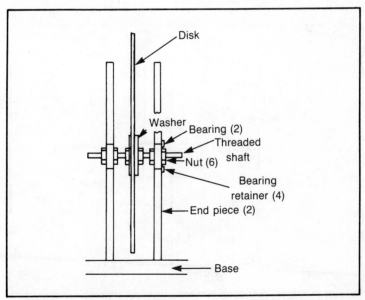

Fig. 3-25. Construction details for the shaft and rotor assembly, using bearings.

labels and paper removed. If desired, a plastic disk about 7 inches in diameter can be used. The record is mounted on a threaded shaft using two washers and two nuts. Tighten the nuts enough to hold the rotor securely, but not enough to break the plastic. The shaft can be supported in the end carries by either bearings or bushings. If bearings are used, they are assembled as shown in Fig. 3-25, with nuts tightened against the inner race of the bearings to hold the shaft securely. The shaft must fit in the bearings without sideways movement. If necessary, increase the shaft diameter using shrink-fit tubing. Typically, the end pieces will be made the same thickness as the bearings, and a bearing retainer attached to each side. The retainer is simply a piece of thin metal secured to the end piece to cover the outer race of the bearing.

The use of bushings is illustrated in Fig. 3-26. The shaft is a brass rod, available in most hobby stores. Obtain a length of rod about 1/8 inch in diameter, and a length of tubing with a 1/8-inch ID. Make sure the rod fits inside the tube with very little sideways movement. Cut a few plastic disks large enough to cover the center hole. Drill a hole for the shaft in each disk, and glue them to the rotor. Make sure they are centered over the record hole, and that both disks hold the shaft "square" with the rotor.

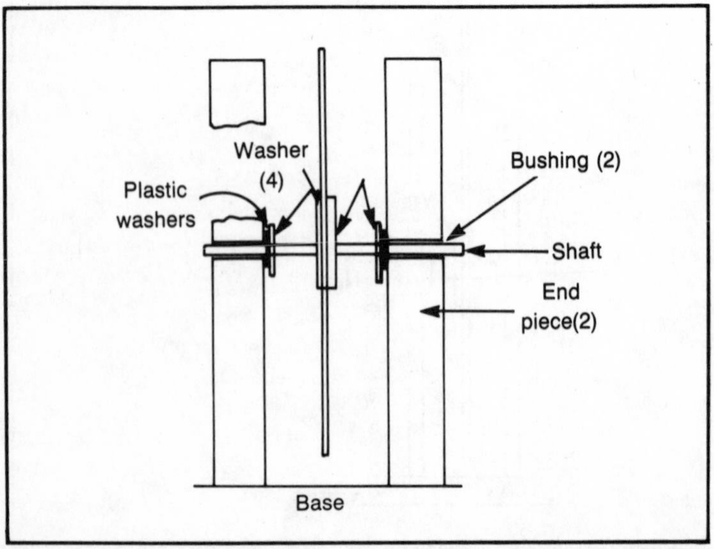

*Fig. 3-26. Construction details for the shaft and rotor assembly, using bushings.*

The shaft is glued to the plastic rotor using good-quality epoxy glue. The outer washers are glued on the shaft with some type of fixture holding them square. Between these washers and the end carrier are two (or more) thin plastic washers on each side. Cut these from smooth plastic, such as a coffee can lid. They act as thrust washers to keep the shaft from moving sidewise.

The bushings are cut from the brass tubing and the ends smoothed down with emery paper. The bushings are glued in the end pieces. Mount the shaft and rotor in the end carriers. If the bearings are used, snug the nuts against the bearings. Do not tighten them too tight, because this will deform the bearing race, causing it to bind. If the bushings are used, mount the end blocks so that there is very little side play in the shaft.

It might be necessary to apply powered graphite to the plastic spacer washers to provide free movement. Spin the rotor; it must spin easy at low speeds, and it must not wobble or move sidewise. See if the rotor is out of balance. Depending how out of balance it is, the rotor might vibrate or just rotate until the heavy side is down. If it stops in the same position, this is an indication that the heavy side is down.

If the rotor has a heavy side, remove a little plastic from the heavy side with a fine file. Do not be overly ambitious and take off too much plastic. It is best to remove the plastic several places around the perimenter than to remove a lot at one place.

When the rotor rolls easily, and is balanced, it is time to make the comb. The comb assembly, as shown in Fig. 3-27, consists of several small sewing needles crimped into a brass carrier. These are not glued because electrical continuity must be maintained, and most glues are insulators. The comb carrier can be either a small brass bar (as shown in the figure) between 1/16 and 1/32 inch thick, or a 1/16-inch diameter brass rod. Brass rod is available at many hobby shops and hardware stores. The sewing needles used as the sharp-pointed needles all must be the same size.

Drill holes in the comb carrier so that the needles fit snugly. Make them about 1/8 to 3/16 inches apart, with a space left for the washer. To position the holes correctly and make them square will probably take a drilling fixture of some type, and some small bits. Hold the carrier in a vise or with a clamp

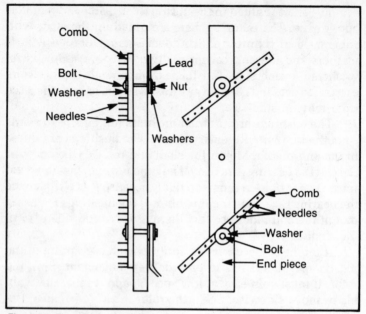

*Fig. 3-27. The comb assembly.*

while drilling. With patience these holes can be drilled.

Mount the comb as shown in Fig. 3-27. Notice that combs are only mounted between the end piece and the rotor. Use a washer to clamp the comb to the end piece. Place a short piece of material the same thickness as the comb between the washer and the end piece, on the opposite side of the bolt. This acts as a spacer, and allows the bolt to be tightened. The comb carrier must be parallel with the rotor, with the needle holes slanted clockwise. To achieve the slant, rotate the comb before tightning the holding bolt.

When the carrier is in place, push the needles through the holes in the carrier so that they are as close as possible to the rotor, without touching. Make sure the needles in both carriers are slanted about the same amount, and in the same direction. It might take some repositioning to get everything correct—when it is, tighten the carrier mount.

Now take a pair of pliers or small vise grips and crimp each needle in place. Note that the leads to the input knobs are also mounted under the nuts on the comb holder. Connect these wires to the knobs mounted on the base (the input terminals). Here again, make sure to remove any clear coating on the knobs. Figure 3-28 shows the assembled motor.

When assembling, be sure to mount the end piece with the combs at the proper distance from the rotor. Some adjustment can be obtained by changing the slant on the needles. When positioned properly, lock into place with screws through the baseplate, and glue. Mount the other end piece so that the shaft holes are aligned, and the shaft does not bind. Attach this piece to the baseplate using screws, and glue.

After the motor is assembled, make sure the rotor is free to rotate without interference. If the needles need to be moved closer to the rotor, place some flat washers between the end piece and the comb. These will act as spacers. Make sure the needles do not touch the rotor, because this will create drag on the rotor, and eliminate the corona action from the needles. Then connect one terminal to some static electricity source, such as a static generator or a Van de Graaff generator, and the other terminal to a good ground. The rotor should start moving, very slowly at first. It might be necessary to gently start the rotor moving, but if all is correct, the static elec-

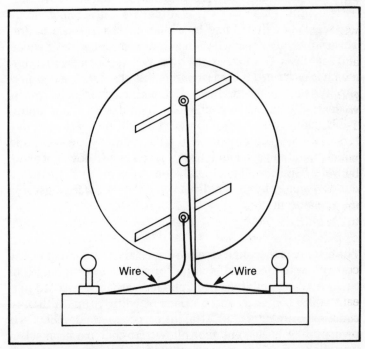

Fig. 3-28. Assembly of the corona motor.

tricity will keep the rotor moving. The problem is what to do when the motor fails to function the first time.

If the static electricity source is present (a test with an electroscope of neon bulb will tell you if it is), make sure the electrical circuit is intact from the source to the comb, and from the other comb to ground. Test the comb by touching it to an electroscope. Then connect the static voltage source to the other terminal and test the other comb for static electricity. Next try adjusting the combs by moving them closer to or further from the rotor. Also, try moving the comb assembly with respect to the radius of the rotor. If the comb is adjusted parallel to the end piece, the direction of rotation is the direction the needles are pointing. If all else fails, adjust the comb in this manner, and gently spin the rotor in the direction of rotation.

Another method of powering the static motor is to use atmospheric electricity, as discussed in the next chapter. There are several other types of electrostatic motors, some of which work better than others. Electrostatic motors, however, have not yet gained the widespread use as that of electromagnetic motors. One reason is that there is a problem generating high enough voltage to power an electrostatic motor. These high voltages can be generated by an electrostatic generator, as described earlier, or by a Van de Graaff generator, which is described it the last part of this chapter. Another possibility is to power the motor using atmospheric electricity, as described in the next chapter, or to design synchronous static motors to be driven from the 60 cycle power Corona motors that have been made with a compound glass rotor and which could run at about 2,000 rpm and have a power of about 90 watts. Granted, this is small compared to the standard induction motor, but it is at least a good start toward a useable electrostatic motor.

### The Electret

The electret is the name given to a special compound of waxes that can hold electrical charges for a considerable length of time. A typical mixture is 45 percent carnauba wax, 10 percent white beeswax, and 45 percent white resin—although some experimenters use straight carnauba wax. When this compound is heated and then allowed to cool in a strong electrostatic field, it becomes electrically polarized. This polarization maintains a positive charge on one surface and a negative charge on the opposite surface. Thus, the polarized wax section is the electrostatic equivalent of a permanent magnet. The electrical charges on the electret may last for a considerable length of time.

Most of the current experimental work in electrostatic motors involves electrets. Many new design motors are being designed and constructed. One motor uses an electret rotor, in which the rotor is made of two sections, separated by an insulator. The sections are reversed, as shown in Fig. 3-29. This rotor is suspended between some charged plates. Because the electret rotor is already charged, there is no need to use brushes or combs to induce more charge to the rotor. Typically, plates or bars are placed so that they just clear the rotor. These are wired to some commutating brushes that maintain the proper charge relationship between the plate and the section of the rotor approaching and the section just past. There must be attraction to the rotor segments just approaching and repulsion to that just past.

Several designs have come forward, but the design parameters and rules have not yet been developed. Most design so far is emperical in nature. These electret motors hold a lot of promise, but need development. Why the wax retains the charge is not understood, and must be investigated. At the present time it is believed that the wax acts as a dielectric in a capacitor, and its molecules become aligned when the wax is hot—then the electrical charges are ''frozen'' into place when the wax solidifies. It could also be that the mobility of the charges is good when the wax is molten, and zero when the wax is solidified. In any case, the parameters that control the electret principle must be developed: What makes a good electret and what makes a poor one? What chemical, mechan-

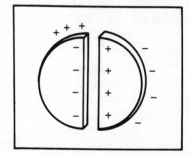

*Fig. 3-29. An electret rotor made from carnauba wax, showing the opposite polarities on opposite faces.*

ical, or process factors influence the quality of the electret?

## THE VAN DE GRAAFF GENERATOR

The modern era of electrostatics began in 1928 with the development of the Van de Graaff generator. It was developed by Robert J. Van de Graaff, a young Rhodes scholar from Oxford University working at Princeton as a National Research Fellow. It was developed as a generator of a constant potential voltage with which to accelerate atomic particles to bombard atomic nuclei in order to obtain information about their internal structure.

The Van de Graaff generator is frequently used in static electricity demonstrations. Typically, it is used to light fluorescent bulbs, create miniature lightning strikes, and provide the energy for Jacob's ladder. The Van de Graaff generator has been used in nuclear laboratories around the world to develop high particle energies required for nuclear accelerators. Energies in excess of eight million volts have been achieved with this generator. Smaller machines have been used for a wide variety of applications.

One of the good features about the Van de Graaff generator is it's simplicity. Small versions, about 12 inches high and costing very little, can develop 50,000 volts. Figure 3-30 shows the schematic of the generator. It consists of a rubber belt driven by an electric motor. Positive charges are introduced onto the belt at the lower end and these are transferred to the upper electrode. The negative charge from the upper electrode is transferred to the lower electrode, and grounded. The electric motor provides motion for the belt, relieving the operator.

Figure 3-31 shows the overall view of the Van de Graaff generator, with the major parts labeled. The Van de Graaff generator provides a continuous source of an electrostatic charge. It is self-sustaining, so that when discharged, the generator will recharge the high-voltage terminal. The operator does not have to continuously turn the crank to generate the voltage, and can use the static charge for demonstrations or experiments.

In operation, frictional contact removes electrons from the moving belt at the driving end and deposits them on a plastic pulley. Positive charges resulting from the lost electrons are carried by the belt to the metal pulley at the upper

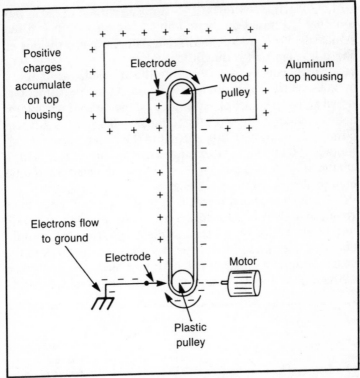

*Fig. 3-30. Schematic of a Van de Graaff generator.*

end. Electrons flow from the metal pulley onto the electron deficient belt, where they are taken to the lower end. As the machine continues to run, heavy charges build up on the both pulleys. After a short time, this charge will reach the ionizing intensity in the vicinity of the electrodes. The electrons are then withdrawn from the upper terminal and sprayed on the belt. There they are moved to the lower terminal. When the charge on the lower pulley reaches the ionizing intensity, the excess electrons will be picked up by the lower electrodes and grounded.

Through this action, the belt continuously exhausts electrons from the upper terminal and discharges them into the earth through the lower electrodes. This leaves the upper terminal with a net positive charge, which, because like charges repel, distributes itself uniformly over the outer surface of the upper terminal.

In theory, the voltage at the upper terminal will increase

without limit. However, in practice, the voltage at the upper terminal is limited by the quality of the insulation and the smoothness of the upper terminal. Any sharp protrusions or sharp corners will form corona points, in which there will be a steady corona discharge into the atmosphere. At about 100,000 volts, the charge leaks as corona from the upper terminal and as conduction current down the insulating column, at a rate equal to about 2 micro-amperes. The belt of this machine is able to carry this current to the upper terminal. Although 100,000 volts is an impressive amount, the machine creates no shock hazard because of the small amount of total charge stored at the upper terminal.

If a well-rounded object is brought to within an inch or so of the high-voltage terminal (the upper terminal), a spark will jump. In this type electrical discharge, the air is rapidly changed from a good insulator to a good conductor, and the spark completely discharges the upper terminal. Reduction of the terminal voltage stops the spark and permits the air to

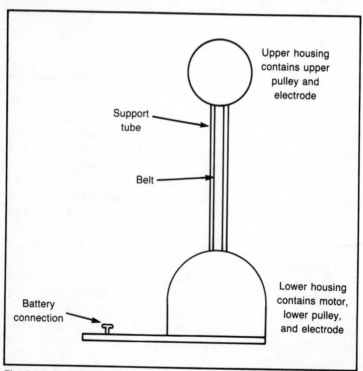

*Fig. 3-31. Overall view of the Van de Graaff generator.*

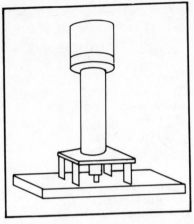

*Fig. 3-32. Assembled Van de Graaff generator.*

regain its insulating strength. So the terminal can be recharged by the belt.

The high-voltage terminal will reach a maximum potential, which it will try to maintain. At this point, the leakage charge will be balanced by the generated charge. If a grounded, sharp-pointed object is brought near the high-voltage terminal, it forms a continuous corona discharge, and decreases the potential. Such a sharp-pointed object, mounted in an adjustable fixture, is often used to maintain a controlled, constant high-voltage terminal potential.

Such a continuous source of high direct-current voltage provides the possibility for many demonstrations and experiments. These range from simple to outstanding demonstrations. First, however, let us look into the construction of a simple Van de Graaff generator.

Figure 3-32 shows the overall view of this machine. Figure 3-33 shows the details of the upper end and Fig. 3-34 shows the details of the lower end. The motor is a small hobby motor, available at most hobby stores. It should be 1 1/2- to 3-volt rating, with decent torque to turn the belt. The belt is a large rubber band, about 3/8 inch wide and 6 inches long. The exact size is not critical, and the rubber belt can be cut from an inner tube. Make sure the rubber belt is clean and dry, and do not handle it any more than required.

The generator is built on a 6- × -10- × -3/8-inch piece of particleboard (or wood or plywood). The motor pulley is a section of 3/8-inch plastic rod, about 5/8 inch long. Drill a center hole in the plastic rod to fit the motor shaft. Mount it

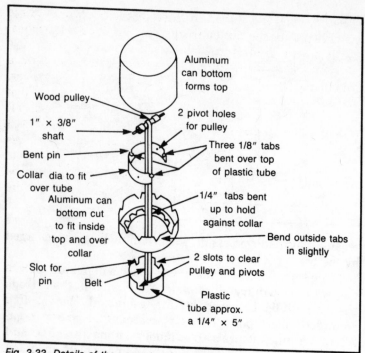

Fig. 3-33. Details of the upper terminal assembly.

on the shaft either by drilling and tapping for a short set screw, or by glueing to the shaft. Groove the center of the pulley to keep the rubber belt in place. Mount the motor on the board so that the pulley is in the center. Cut a 3-inch square plywood platform. Cut a hole in the center slightly smaller than the plastic tube. Cement the plywood platform with the hole centered over the motor pulley with some wood dowels at each corner. The 3/8-inch dowels are long enough to clear the motor housing. Connect the motor leads to terminal clips mounted on the board.

The length of the plastic tube is determined by the length of the rubber belt. The belt, when in position over the pulleys, should not be taut, but should fit over the two pulleys in a straight line. It is advisable to have some spare belts of the same length on hand in case the belt is damaged and needs replacement. Use a 1 1/4-inch diameter plastic tube a little longer than required, the tube can be cut to size during final assembly. For the 6-inch rubber band, use approximately 5 inches. The bottom end of the tube is cut flat and square, while the upper end is notched in three places as shown. These notches are for the pulley pivots and the upper electrodes. These are cut later, during the assembly.

For the upper terminal use an aluminum can with a rounded bottom. It must be at least 2 1/2 inches in diameter, and not have any ridges or protrusions. It has been found that a 12-ounce beer can with a rounded bottom works well, although some soda cans will work; make sure they are at least 2 1/2 inches in diameter. Also make sure there are no ridges or sharp seams on the bottom or sides. These will dissipate the charge. If there are some sharp projections, they can be removed with a file. Remove all paint from the outside of the can, and make sure that the inside is clean and dry.

Cut one 2-inch section from the bottom of the can, as shown. Fabricate the aluminum collar of heavy-gauge sheet aluminum to fit around the outside of the plastic tube. Cut the three tabs and bend them as shown to keep the collar in place. Using another aluminum can bottom of the same size, cut the bottom piece as shown. The inner set of tabs bind against the collar while the outer tabs fit inside the upper portion. Figure 3-33 shows the details of the pieces, and Fig. 3-34

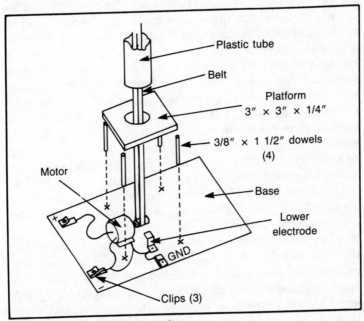

*Fig. 3-34. Details of the lower portion.*

shows how they fit together. Cut out the bottom section center hole and bend the tabs so the bottom section will fit snugly over the aluminum collar. Cut out the top tabs and bend them in a little so that the top section will fit snugly over the bottom section.

Make the upper wood pulley as shown. Groove and round the center section to keep the rubber belt from slipping to one side. This can be accomplished with a woodworking lathe, with a sharp knife and some sandpaper, or by turning it with a drill. The shaft is made from some brass rod, about 1/16 inch in diameter, pointed at the ends. The points fit through the two holes in the collar, so they must be smooth.

Make sure the shaft is inserted into the center of the roller so it will rotate freely without wobbling. One way to do this is to finish sanding the roller by turning the shaft in a drill. This will assist in keeping the roller round. The points become the pivot points for the upper roller, so the points and the holes must be smooth, with no burrs. The holes for the shaft are sized so that the points fit only part way through. The shaft must be of the correct length so that the shaft is turning on the points.

Drill a hole in the collar for the pin. This hole should be slightly smaller than the pin, so that it is a forced fit. This pin is mounted after the collar is in place around the plastic tube. This pin becomes the upper electrode. Place the wood pulley, and shaft, complete with rubber belt, in the holes in the collar. Position the plastic tube over the hole in the lower platform. Place the collar over the plastic tube, making sure it is all the way down. Put the rubber belt over the plastic motor pulley, making sure it is not twisted. Make sure the pulleys are free to turn, and that the belt is not stretched. It should be tight enough to prevent the belt from moving to one side or the other, and loose enough not to cause undue drag on the shaft and pivots. If the belt is too tight, cut a little off the bottom of the plastic tube. If the belt is too loose, add some wooden pieces under the lower end of the tube.

The wooden pieces must have the same size hole, and glued over the hole in the existing bottom platform. The two pulleys must be parallel, so that the belt will not "walk" to one side or the other of the pulleys, or completely off the pulley. The best way to test this is to run the motor, and watch to see if the belt works one way or the other. If it does, lift

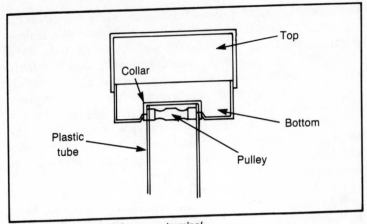

*Fig. 3-35. Assembly of the upper terminal.*

one of the tabs on the collar to adjust the upper pulley. When the correct tab, and amount have been determined, bend the tabs. Run the motor some more to assure proper alignment. If worse comes to worse, the tabs on the collar might need to be recut, or new pivot holes drilled for the shaft.

When this is all squared away, push a pin into the pin hole in the collar. The pin point should be toward the belt, and be as close as possible without touching the belt. Hole the pin with some pliers, and bend it up, over the top, and back down the inside of the collar. This will "lock" it into place and keep it from moving. The pin should be on the side of the belt that is moving upwards.

Most dc motors are reversible by reversing which input the positive voltage source is connected to. Determine the correct polarity and attach the motor leads to the clips on the base. Park the polarity of each clip. Solder a needle or pin to a 1/2-inch metal bracket, and position it as shown for the lower electrode. Again, the pin point should be close to, but not touching, the rubber belt. Fasten the bracket to the base with a solder leg connected to a ground clip. A lead should also run to the motor housing and to the negative motor voltage clip, as shown.

After the pulleys have been adjusted, and the upper electrode secured, it is time to finish assembling the top. Figure 3-35 shows the top assembly. The collar is in place against the plastic tube. Carefully slide the lower section over the pulley to the position shown. The inner tabs support this section against the collar. Bend the tabs to assure a good, solid connection. Bend the upper tabs in very slightly. While flexing the upper tabs in, slide the top section over the tabs. Make sure the upper section is seated securely. This assembly should be self-supporting, being held in place by the collar.

Once again, check the rubber belt position and tightness. When everything checks out alright, glue the plastic tube to the platform. It is advisable to provide some positioning blocks on the platform that keep the plastic tube in place. This will help hold the bottom of the tube in place. Make sure the belt and pulley assembly turns freely.

Make up the test lead shown in Fig. 3-36. This test lead will facilitate testing of the generator, and be used for some of the demonstrations. It consists of a 24-inch length of wire connected to a knob. Crimp a solderless ring terminal on the wire, and secure this to the wire with a bolt in the knob mounting hole. Use a wooden handle to hold the test lead. A spring-type clothespin can be used for the handle. Using fine emery paper, remove any clear coating on the metal knob. Connect the other end of the wire to the ground test terminal on the base.

This makes a grounded metal ball that can be used to test the unit. The bolt is long enough to secure to the knob, for the nut and for the wooden clothespin to grip. Run the nut up tight against the knob, to secure the ring terminal to the knob. Wrap the bolt and top part of the clothespin with electrical tape. The Van de Graaff generator should be ready for the first demonstration.

Connect the battery to the battery terminals, positive to the positive terminal. The unit should start operating and the

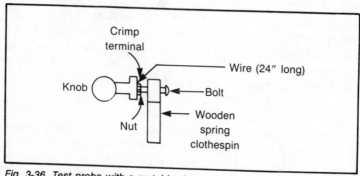

Fig. 3-36. Test probe with a metal knob.

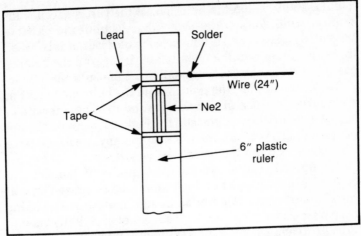

Fig. 3-37. A Ne2 neon lamp test probe.

belt should be moving. Let the unit run for a minute or so to let the charge build up. Then hold the test lead by the wooden handle, and bring it close to the upper portion of the generator. There should be sparks between the upper terminal and the metal knob. Gradually move the metal ball outward, away from the upper terminal. When the sparks stop, the limit has been reached.

Move the knob back toward the terminal unit the sparks barely start. This is the maximum distance that the sparks will jump. This should be done in a darkened room to make even the weakest sparks visible. When the distance to start drawing a spark is determined, measure it with a ruler. Then, using a pump atomizer, spray a fine mist of water in the area and try to draw a spark again. Notice the difference in the distance which a spark can be sustained. A Ne2 neon lamp can be used.

Solder a wire to one of the leads and tape the lamp to a wood or plastic handle. A 6-inch plastic ruler work fine. This test probe is shown in Fig. 3-37. One lead serves as the input lead, and is bent straight out. The second lead is soldered to a 24-inch length of wire. The Ne2 is taped to the 6-inch ruler with some narrow pieces of tape, cut to about 1/4 inch wide. This prevents the tape from blocking the Ne2 illumination. One piece of tape goes around the leads next to the tube and the other piece goes around the opposite end of the tube. For best results, use a dark-colored ruler as the handle, and position the bulb so that the two electrodes inside the bulb are side by side. The electrodes are visible through the glass, and are connected to the leads. It is between these electrodes that the neon tube "fires," and the colored glow is seen. In a darkened room, bring the bulb close to the upper terminal. Find the point where the bulb just turns on, and move the probe in and out, around the upper terminal. In this manner, the electrostatic field intensity around the upper terminal is determined. Look for any weak spots or "holes" where the intensity goes way down.

Support a sewing needle by an insulating handle, and connect it to the upper terminal. Molecules of ionized air will rush from the needle point as though they were streaming from a jet under pressure. This stream of air will be enough to cause a candle flame to flicker, if not extinguish completely. The stream of ionized air may be visible in a dark room. Hold the back of the needle against the upper terminal, as shown in Fig. 3-38. This shows a needle taped to a stick. Any insulating handle can be used, such as a plastic drink swizzle stick, or a popsicle stick.

The needle is held in contact with the top terminal. It might need to be laid across the terminal to achieve maximum contact between the needle and the terminal. This "electric wind" can be made to turn a small vane. From aluminum foil, fashion a "spinner" with 4 or 6 blades. Use a straight pin as a shaft to support the spinner and a wooden stick as the handle. This is similar to the wind spinners available at the store. In a darkened room, it may be possible to see a corona or St. Elmo's fire, emanating from the point of the needle.

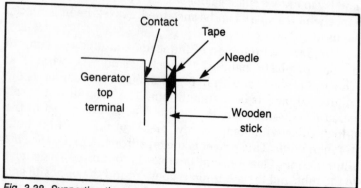

Fig. 3-38. Supporting the needle against the upper terminal.

Fluorescent lamps will light up when touched to the high voltage terminal. In a darkened room, filament-type lamps will glow various colors, depending on the type of gas they contain. These must be in the clear glass type, preferably flashlight bulbs. Even some old radio tubes can be made to glow by touching them to the high-voltage terminal.

A miniature aurora borealis can be made by boiling a small amount of water in a thin flask. When all the air has been displaced by the steam, stopper the flask and allow it to cool. After the steam has condensed, a partial vacuum is formed. There will be some water vapor, and no air. This rarified combination will glow greenish and pink when the flask is touched to the high-voltage terminal. As long as the operator is not fully insulated from ground, there will be no charge buildup on the operator. But if the operator is wearing insulated shoes, such as these with soft rubber soles, and is not touching ground directly or indirectly, there is the chance of a charge buildup. This is not dangerous, except that there can be a spark discharge when touching ground, which can be slightly painful—or just startling. It is best not to try to demonstrate this, because we know the results. It is the same as the static discharge when walking across a nylon rug. There is no need to receive additional shocks of electricity.

An amusing demonstration that can be done is the "jumping ball" demonstration, wherein some small balls jump around inside a closed tube. Make up six small balls from pith or similar material, such as styrofoam. Coat these balls with a conductive material, such as soot or graphite. Make a cage, or tube, from a strip of transparent plastic, rolled into a cylinder and capped with metallic caps, such as those from peanut butter jars. Connect the caps to the Van de Graaff generator terminals: One to the high-voltage terminal and the other to ground. As power is applied, the balls will be attracted to one of the lids. The balls will give up their charge and receive the opposite charge. The balls will be attracted to the opposite lid. This will go on as long as power is applied. This works on the principle that opposite charges attract, and like charges repel.

Figure 3-39 illustrates this device, and how it works. The pith balls are covered with a conductive material to allow the transfer of charge from the metal caps to the pith balls. These balls are light enough to be bounced up by the electric charge.

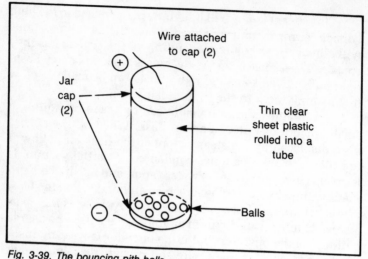

*Fig. 3-39. The bouncing pith balls.*

The jar lids that form the ends of the tube are, in reality, extensions of the terminals of the Van de Graaff generator.

The system starts with the balls laying against the bottom, or negative terminal. Because it is the negative terminal, electrons flow from the ground to the balls. When enough electrons have accumulated on the ball, the force of repulsion between the lower terminal and the balls will cause the balls to move upwards. This is assisted by the opposite charge (positive charge) on the upper electrodes. When the pith ball touches the upper lid, the excess electrons are conducted to the positive terminal. When the electrons are removed, there will be a point at which the charge will no longer support the pith ball. Then the ball will fall to the bottom, or negative terminal, and the process will repeat.

During Ben Franklin's investigation and experimentation with electricity, he found that a sharply pointed, grounded metal rod would draw off the electrical charge without a spark discharge. This led to the invention of the lightning rod, which is a sharp-pointed, grounded metal rod, placed in the highest area of the building to be protected. The next project demonstrates the idea.

First, assemble a plastic experiment table, as shown in Fig. 3-40. This table is used as an insulated surface on which to conduct experiments. It consists of an 8-inch square of 1/8-inch plastic, mounted on four legs of 1-inch (diameter)

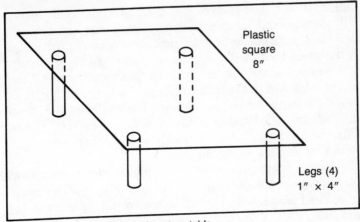

*Fig. 3-40. The insulated experimenters table.*

by 4-inch (long) pieces of plastic rod or wood dowels. On this mount a 1 3/4-inch-high section cut from an aluminum can with all finish removed, as shown in Fig. 3-41. Cut a tab in the can, and attach a ground lead to it. Secure the ground lead to a ground, or to the third wire in the ac receptable. (Note: be careful when around ac receptables, and use only the third wire, or the round terminal, as ground.) Whenever the Van de Graaff generator is in use, the ground terminal must be connected to some good ground. This prevents charge buildup, and the possibility of an electrical shock.

Bring the operating Van de Graaff generator close to the

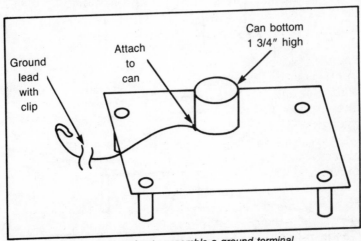

*Fig. 3-41. Using a can section to assemble a ground terminal.*

unit, as shown in the top of Fig. 3-42. At some point, sparks will be observed jumping between the metal can and the generator. Note the farthest distance that the sparks will jump. A darkened room will improve the viewing conditions.

Attach a small lightning rod to the top of the metal terminal as shown in the bottom of Fig. 3-42. Tape a length of No. 22 bare, solid copper wire to the center of the metal ground terminal. Bend up 3/4 inch of the free end of the wire. Clip the end of the wire diagonally so that it is pointed. The wire represents a lightning rod.

Move the operating generator to places where good sparks were observed before. There is no spark discharge. The sharply pointed wire has drawn off the electrostatic energy by corona discharge. This corona discharge is visible in a darkened room. The charge is dissipated from the generator, picked up by the "lightning rod" and grounded. The sharp point results in an extremely high electrostatic charge concentration at the

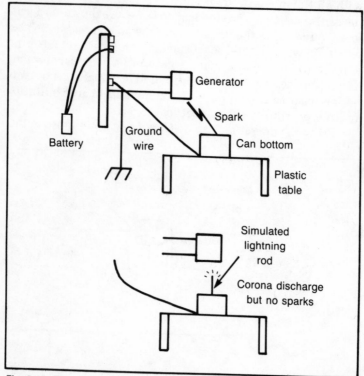

*Fig. 3-42. The simulated lightning demonstration.*

point. This will pick up the charge from the generator, restricting the maximum charge on the upper terminal. It acts as a maximum voltage regulator, and it can be varied by changing the distance between the point and the terminal.

The Van de Graaff generator will power the electrostatic motor. Connect the negative input to the electrostatic motor to the generator ground. Then touch the upper terminal of the generator to the positive motor input. The motor should run, but might need to be gently started in the proper direction. Note that the generator does not have to be on a flat surface to operate. It can be held by the base board and tilted, or turned completely upside down, and it will still remain working.

There are some important pieces of information about Van de Graaff generators that should be passed along. First, approximately 50 square inches of belt per second passing over the pulleys will produce one microampere of current. This means that the wider the belt, the more the charge. It also follows that the faster the belt runs, the greater the charge, but this is valid only up to a point, after which the results decrease.

The maximum feasible speed is 100 feet per second, but this requires bearings for the upper pulley. It might increase the charge produced by running the motor with a higher voltage battery. For instance, if the motor is a 1 1/2-volt battery, try running it with 3 volts. If this is done, keep a close check on the motor, so that it doesn't overheat. Also make sure the belt is running faster with the increased voltage, and that not all the additional power is lost to friction. Along the same line of thought, a wider belt will produce more charge. If required, cut a band from an inner tube. Two ends of a piece of rubber can be glued together forming a belt. Overlap the ends at least 1 inch, and cement them together using good-quality rubber cement or contact glue. When the glues has thoroughly dried, taper the ends so there is no thicker section.

If the belt has been made wider, the electrodes need to be made wider. There are at least two ways to do this. First, additional needles can be mounted in parallel with the original electrodes. This requires additional needles mounted on the same brackets, and could be a difficult problem. An alternative is to use a section of metal screen the same width as the belt. This screen can be attached to the brackets that held the needles (Fig. 3-43). Make sure the screen is metal, and that there are bare wires pointing to the belt, as shown. Short tufts, or bunches, of small, bare wire can be soldered to the brackets. This forms an electrode like a wire brush, with a bunch of bare wires pointing toward the belt.

One of the limiting factors as to the maximum voltage on the upper terminal, or collector, is the smallest radius of curvature of the collector, in inches. The maximum voltage is about 70,000 times the smallest radius of curvature of the collector, in inches. Thus, a perfect sphere 12 inches in diameter will have a theoretical limiting potential of 420,000 volts. Holes in the sphere for the plastic tube alter the field pattern and reduce the maximum.

The most effective compromise for the ideal shape consists of a spheroid slightly flattened at the bottom. The minimum radius of curvature should be located at a reasonable distance from the insulator to discourage sparking along the insulator surfaces. To realize a large fraction of the theoretical maximum voltage, the collector must be at least two or three diameters removed from other metallic parts. The distance should be greater if sharp-edged metal parts are present. Covering all parts by a rounded metal shield of large radius of curvature helps. The opening that admits the belt should not be much larger than half the diameter of the collector, and should curve inward.

Therefore, to improve the operation of the above Van de Graaff generator, do four things: First, enclose the bottom elec-

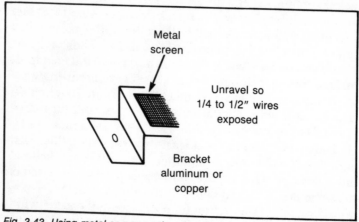

*Fig. 3-43. Using metal screen as the electrodes.*

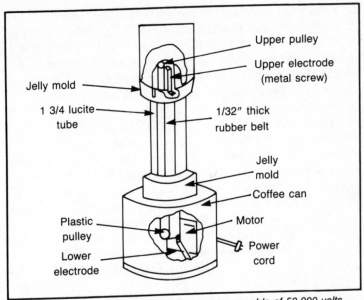

*Fig. 3-44. An improved Van de Graaff generator capable of 50,000 volts.*

trode and motor in a metal housing. Then add a wider belt, making sure it will run on the pulleys. Next, change the electrodes from the single pin type to a multiple pin or screen type electrode. Finally, replace the top terminal with a metal sphere. This will improve the system to the maximum.

If so desired, an improved Van de Graaff generator can be assembled. Figure 3-44 shows a Van de Graaff generator capable of 50,000 volts. It is driven by an ac motor, like that in a phonograph. It uses metal screens as the electrodes, and small ball bearings or plastic sleeve bearings for the upper pulley. The belt is made from an innertube. The upper terminal is made from a jelly mold, and a rounded bottom can.

Support the electrode by a plastic collar around the Lucite tubing, or by tabs on the inner part of the upper terminal. Mount the upper electrode on a short strip of metal, attached to the inside part of the upper electrodes. No matter which method is used to attach the upper electrodes, the area of the lower portion of the upper terminal should be curved upward. This can be done with tabs, similar to previous example.

It is also possible to find a metal container or pan in the proper shape, complete with a raised center section. A similar jelly mold is also used as the top part of the lower section. This mounts on an inverted can with a snug-fitting lid. This metal can serve as the lower housing, containing the motor and the lower electrodes. Mount these inside the can body, with the lid secured to the base of the generator. It might be necessary to provide some means of locking the can into the can bottom.

The tube is Lucite, about 9 inches long, mounted in the lower section. Cut the tube at the top as shown in Fig. 3-45, and mount the bearings in the sides of the tube. This figure also shows a flange for the upper terminal to rest on. The upper pulley is made from brass, and is about 3/4 inch in diameter. Mount it to the shaft with a set screw, and secure the shaft in the two bearings. The roller should roll freely in the bearings. The lower bearing is cut from 3/4-inch polyethylene, and is mounted directly on the motor shaft.

The motor must be mounted so that the pulley is in the proper place, directly below the tube. The belt can be made from 1-inch-wide strip of gum rubber. An alternative is to use an old inner tube, and cut a 1-inch loop the proper length. If required, cut a 1-inch strip and glue the ends as detailed previously. The belt should be just tight enough to stay on the pulleys, but not tight enough to impose undue load on the motor.

Because this unit is powered by a 110-volt motor instead of a battery-driven motor, care must be taken and the wiring done in a neat and safe manner. Mount a 110-volt on/off switch on the base. This should be a double pole, single throw

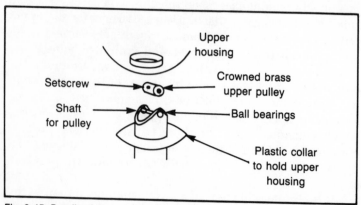

Fig. 3-45. Details of the upper end of the tube for the Van de Graaff generator.

switch, so that both sides of the 110-Vac line is opened. The third wire of the ac line cord can go to the ground in most homes and buildings wired with the three wire outlets. Add a fuse between the switch and the motor. This will protect the power systems, and the operator, in event of a short or problem.

This Van de Graaff generator is no toy, it will generate several thousand volts. It is an instrument for a serious experimenter, and as such must be treated with respect. Remember that the generator is powered from the 110-volt power line. This voltage is dangerous, so care must be taken. Allow no exposed bare ac power line. Be careful not to get anything caught in the belt. Make sure the motor is securely mounted inside the lower metal container.

The electrodes are made from either metal screen or short pieces of small bare wire. Mount these on small brackets, and then on the metal terminals. They are both on the side of the belt that moves from the lower to the upper terminal, and should be mounted as close as possible without actually touching the belt.

There is no course of study dealing mainly with the interesting and intriguing field of statics and the related fields. It is mainly taught in association with physics, or a similar subject. Many of the principles of statics are empirical, with no fixed or firm theory to justify the results. There is considerable room for new developments in the area of electrostatic motors. These developments will come from the small, private experimenters working either in home workshops or in universities for advanced degrees. Until more developments are made that give a distinct commercial possibility to the electrostatic motor, industry will not enter this area in a large way. So it is up to interested individuals.

# Chapter 4

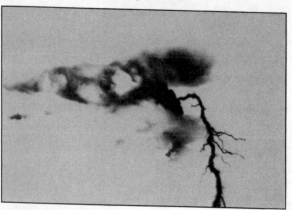

# Atmospheric Electricity

T HE GREATEST SOURCE OF NATURALLY PRODUCED ELECTRICITY is in the atmosphere. The atmosphere is made up of several different zones, as shown in Fig. 4-1. The dividing lines between the regions are not well-defined lines, but gradually changing properties. The altitudes given are for reference; the properties at any given point are subject to change, depending on temperature, pressures, and other factors. The northern lights, or aurora borealis, occur in the stratosphere.

One clear sunny morning, a ham radio operator started to hook a ground wire to a 50-foot antenna. A fat spark jumped from the metal tower to his hand, and knocked him off the roof. Fortunately, he landed in a stack of straw. The weather was hot and dry, with no breeze of any consequence.

A television viewer on a Pennsylvania farm was plagued with strange electrical disturbances that obscured the TV picture. He found small sparks jumping from a metal cupola on his house. Grounding the cupola ended the TV problems.

In these two, and in many other instances, sparks have been produced where there is apparently no source. These sparks are an indication of the amount of electrical energy present in the atmosphere. The atmosphere contains several different generators of electricity, such as the Van Allen belt, cosmic rays, thunderstorms, lightning, and static electricity.

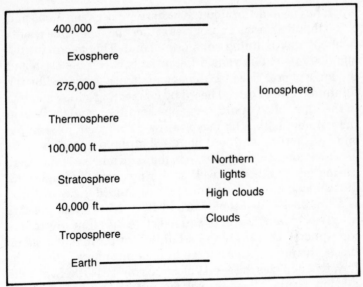

*Fig. 4-1. The zones, or divisions of the atmosphere.*

These result in a voltage gradient of several thousands of volts, with several hundred amp of current. Electrical current is the number of electrons per unit time. It is a measure of the amount of charge flowing past a point. One amp of current is $628 \times 10^{16}$ electrons per second flowing past a point. This is 628 with 16 zeros. This is quite a few electrons in one second.

## THE ATMOSPHERIC PROFILE

The earth is like a big battery, continuously losing electrons to the atmosphere. Although air is a poor conductor of electricity, it can, nevertheless, conduct electricity away from the earth. It has been estimated that the earth would lose almost all of its charge in less than an hour unless the supply were continuously replenished.

Most scientists agree that the steady loss of electrons to the atmosphere is balanced by the thousands of daily thunderstorms that pump electrons back into the earth. All clouds modify the flow of electricity, but most of the replenishment is from the thunderstorms and lightning strikes. About four out of every five lightning strokes that reach the ground transport electrons back to the earth.

It has been determined that there are between 2,000 and 6,000 thunderstorms in progress at any one time. This results in about 100 lightning bolts per second. This maintains the 300,000 to 400,000 volts differential between the earth and the ionosphere. The electromagnetic noise generated by the lightning strokes can be heard by connecting a long antenna to the input of an audio frequency amplifier, and turning up the volume. Lightning makes a sharp, buzzing noise in the audio spectrum, and will be heard at the output.

Electron flow exists between the earth and the atmosphere during fair weather, as well as during periods of lightning or thunderstorms. This condition is illustrated in Fig. 4-2, and is sometimes called ionic current flow. Notice that positive ions flow toward the earth and negative ions flow toward the ionosphere. This results in a buildup of positive charge on the earth and negative charge in the ionosphere. This is the flow that depletes the excess electrons on the earth. Even with this ionic current flow, the voltage potential between the earth and the ionosphere is about 300,000 volts. It is this voltage difference that powers the current flow illustrated in Fig. 4-2.

The earth's surface is the negative potential while the ion-

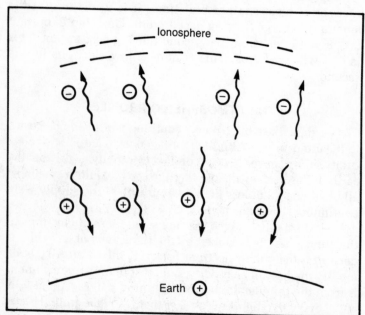

Fig. 4-2. The fair weather current flow in the atmosphere.

osphere is the positive potential. But, because air is not a perfect insulator, and can conduct electrons, there is current flow. In fact, air is a relatively poor insulator, with typically 200 ohms between the earth and the ionosphere.

As illustrated, this current flow consists of positive ions flowing toward the earth, and negative ions flowing the opposite direction. The rate of flow has been calculated to be between 1,400 and 1,800 amperes. Because it originates from the entire surface of the planet, the current is too small for us to feel, being about 9 microamps per square mile. Sensitive measurements have shown that this estimate is reasonably accurate.

Because there is a voltage difference between the earth and the ionosphere, there is a voltage gradient set up between the earth and the ionosphere. Due to the changing conditions in the atmosphere with height, this gradient is not uniform. Near the earth, it is slightly less than 4 volts per inch. This gives a difference of about 260 volts between the feet and the head of a 6-foot man.

A voltage, or potential gradient, is the change in voltage per unit length. This can be linear or non-linear. If it is linear, the gradient is determined by dividing the total voltage by the distance separating the positive and the negative terminals. For a non-linear gradient, the voltage change is not constant per unit length. Figure 4-3 illustrates both a linear and a non-linear gradient. The gradient for the linear portion is 10 volts per inch, while the gradient for the non-linear is a function of where the reading is taken. At the far left it is 10 volts per inch while at the right side it is 300 volts per inch. This is the concept of voltage gradients.

Figure 4-4 shows the circuit equivalent of the atmosphere. The equivalent resistance is 200 ohms, and the capacitance is 0.25 farads, or 250,000 microfarads. Since the atmosphere is acting like a large, low-resistance capacator, the charge stored is a function of the voltage and capacitance. The equation is:

$$Q = CV$$

Where:

> Q is the charge in coulombs
> C is the capacitance in farads
> V is volts

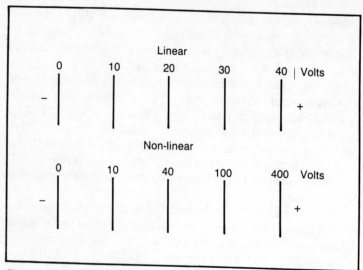

Fig. 4-3. Linear and non-linear voltage gradients.

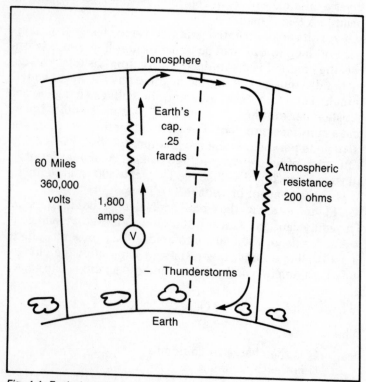

Fig. 4-4. Equivalent circuit of the atmosphere.

Calculating this out:

$$Q = .25 \text{ farads} \times 300{,}000 \text{ volts} = 75{,}000 \text{ coulombs}$$

This atmospheric profile will be discussed in greater detail when discussing the profile of a thunderstorm. It is thunderstorms that create a major discontinuity in the otherwise smooth profile.

The normal fair-weather potential gradient varies from month to month, with January being the highest—about 20 percent above normal. This is because the earth is closest to the sun in January. So July, when the earth is the farthest from the sun, is the lowest—about 20 percent below normal.

Falling raindrops often transport positive ions back to earth, leaving the electrons in the atmosphere. So, precipitation contributes to the loss of electrons from the earth by leaving them in the atmosphere during the evaporation/precipitation cycle. But the prime carrier of electrical charge during fair weather is the atmosphere's ions. These ions transport electrons by passing electrons from one ion to another, or by the actual movement of negative ions. The ions, some positively charged and some negatively charged, are in constant motion. So, in a random manner, the ions may collide with each other. Some are restored to their unchanged state by picking up or losing electrons. But new ions are always being formed by radiation or collisions. All of the molecules of the atmosphere can become ions, so some of the ions, like hydrogen, are "small" and some, like carbon dioxide, are "large." The small ions are lighter and more mobile than the large, heavy ions.

Benjamin Franklin was one of the first to be intrigued by the ability of sharp points to draw electricity. He observed that a charged body with a sharp point or rough edges lost its charge faster than a flat or smooth body. In nature there are many points: blades of grass, leaves, or edges of sand, for example. These natural points conduct electrons from the earth and discharge them into the air. When the ions collide in a concentrated area, such as the charged region of the tip of a point, additional ions are produced and a transfer of electrons takes place between the ions and the point. This is called point discharge. Figure 4-5 illustrates point discharge in a strong potential gradient beneath a thundercloud.

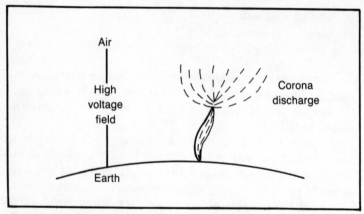

*Fig. 4-5. Point discharge, or corona, from a plant point.*

The electron flow from a point is normally a quiet, invisible process. However, if the potential gradient becomes sufficiently great, the electrons leave the point with enough energy to ionize the air around the point, causing the air to become luminous. This visible ionization is often called corona discharge. It is also called St. Elmo's fire. This ionization has been often mistaken for fire, and will behave somewhat like flames. The color of the ionization is a function of the energy of the ions, and the elements in the air that are ionized. Different materials require different energy levels to ionize, and, consequently, result in different colors when ionized.

The amount of color is a function of the number of ions formed. Where there are very few ions formed, the colors will not be seen. The color will always appear transparent and in motion at all times. When the color appears solid and starts to glow, it becomes plasma.

## WHERE ATMOSPHERIC ELECTRICITY COMES FROM

If all this electrical energy is present in the atmosphere, some natural phenomena must be responsible. In fact, there are a few responsible, and the major ones are: solar wind, cosmic rays, natural radioactive decay, static electricity, electromagnetic generation.

Although it is difficult to determine the contribution each method makes to the overall picture, they all do contribute. The solar wind is the stream of high-energy particles emanating from the sun and permeating the Solar System. When these high-energy particles collide with atoms in the upper atmosphere, ions are formed, contributing to the total charge.

Cosmic rays and the solar wind are the same thing: high-energy particles in space. But the solar wind is considered to be generated at the sun, while cosmic rays exist throughout space. These cosmic rays reach and penetrate the envelope of air that surrounds the earth, so most of the energy is spent before they reach the earth's surface—although enough energy is left to cause sunburn and excess heat. The atmosphere shields us from the high-energy radiation of the cosmic rays. In dissipating their energy while passing through the upper layers of air, the cosmic particles collide with air molecules, creating ions.

Not all the ions in the atmosphere are produced by external energy sources. Some are produced by internal sources. Natural radioactive disintegration of radioactive materials, ionizes some of the air near the surface. Two additional natural sources are the electrostatic interaction between moving air and the earth, and the generator effect of the wind moving air molecules across the magnetic lines of the earth's magnetic field. The amount of contribution these last two sources make is very small, while the solar wind, or cosmic rays, account for most of the atmospheric ions.

The ions within the atmosphere can be ions of any of the elements that make up air. Most of the ions will be oxygen and nitrogen because these elements make up most of the air. The ion concentration near the ground is about 65,000 ions per cubic inch. This is infinitesimal when compared to the billions of neutral molecules in the same cubic inch. The concentration of ions, as compared to the total number of molecules increases as the altitude increases. At about 40 miles up, there are more ions than neutral particles. There are 10 times as many electrons at 180 miles up than there are at 70 miles up.

Above the ionosphere is the exosphere. It is in this region that the Van Allen radiation belts exist. These radiation belts, or zones, were named after Dr. James Van Allen of the University of Iowa. He directed some of the satellite and rocket experiments that studied this zone in detail. These are belts of highly charged particles that have apparently become caught by the magnetic field of the earth. These radiation belts surround the earth and are the most dense above the equator and weakest around the poles. Figure 4-6 shows the Van Allen belts encircling the earth. They are affected by the solar wind, which tends to drag the belt several hundred miles into space. The belts behave in mysterious ways over the poles, coming down almost to the earth.

There is much yet to be learned about these belts. For instance, was it the nuclear explosions that generated the Van Allen belts, or is it a natural phenomena? Most agree that they are natural, but they are affected by activity on the earth.

The effect of the Van Allen belts upon the electrical balance of the earth is not known. It seems that the interchange of electricity between the earth and the atmosphere is controlled primarily by the umbrella of the ionosphere that encloses the electrical process—of which the thunderstorm is a major part.

## PROFILE OF A THUNDERSTORM

A thunderstorm disrupts the even movement of ions, and adds a large discontinuity. The profile of a thunderstorm is shown in Fig. 4-7. There is a separation of charge in the upper and lower regions of the thunderstorm. The upper regions are predominant positive while the lower, or base regions are

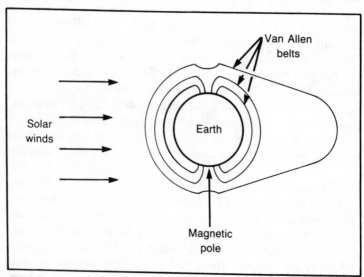

*Fig. 4-6. The Van Allen belts encircling the earth.*

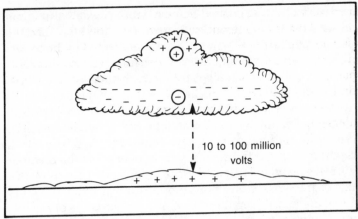

*Fig. 4-7. Profile of a thundercloud, showing the charge distribution.*

negative. The 10 to 100 million volts is the voltage differential between the ground and the base of the thundercloud. This has been tested, but not measured, by flying a kite into the lower regions of the thundercloud. A good spark will jump from the kite string to ground. Do not hold on to the string because when it becomes damp it is a good conductor.

In fact, more than one case of intense point discharge has been observed from the frazzled strands of a kite wire, in clear weather. This is a result of the voltage gradient as discussed previously.

The distribution of the charges within a thundercloud is, at least partially, due to the violent air movement within the clouds during the early stages. Figure 4-8 illustrates these movements in a just-forming thundercloud. An updraft drives the warm, moist air past the condensation levels. The vertical deployment is aided by air entering the cloud from the sides. Heat energy is released by the condensing of water vapor and the outflow of air aloft. The second stage in the thundercloud formation starts when the rain and snow begins to fall within the cloud. The charge separation occurs with the updrafts and the water vapor condensation, and continues through the early portions of the second stage.

Updrafts of warm, moist air, rising into cold air can cause small cumulus clouds to grow into the large cumulonimbus cloud systems associated with thunderstorms. These turbulent cloud systems dominate the atmospheric circulation and electrical field over a wide area. It takes as little as 30 minutes for the small cloud to grow into the turbulent, electrified giant.

As the thundercloud develops, rain and snow begin to fall within the cloud, and an ice phase appears in the towers, or the extreme upper portions. These towers continue to grow upward and may intercept the high-altitude winds. During this phase, a downdraft joins the updraft, and the charge distribution is set. At this time lightning and heavy surface rains develop. Eventually the downdraft destroys the thundercloud. As the surface winds shift from convergent to divergent, the updraft is cut off from the source of energy, so the precipitation weakens to a light surface rain, and soon stops. The downdraft ceases and the thunderstorm is over.

As the thunderstorm develops, interactions of ions, external and internal electric fields, and complex energy exchanges produce a large electrical field within the thunderstorm. No completely acceptable theory has been developed to completely explain the complex process of thunderstorm electrification. But it is believed that the electrical charge is important to the formation of raindrops and ice crystals, and the electrification closely follows the precipitation.

The distribution of charge is usually a concentration of positive charges in the frozen upper layers, and a large nega-

Fig. 4-8. Profile of a thundercloud, showing the movement of air and the temperatures.

tive charge around a positive area in the lower portions of the cloud.

The rain and snow that is falling within the cloud make it to the ground, usually in the form of rain. This heavy surface rain is accompanied by lightning within the cloud and from the cloud to ground. During this phase in the life of a thundercloud, the air movement has changed from the original updraft to a downdraft, pulling the moisture from the upper parts of the cloud to the lower portion, where it feeds the surface rains.

The thundercloud interrupts the fair weather profile of the atmosphere by adding a path for the electrons to be returned to the earth. Figure 4-9 shows the charge movement cycle for the fair-weather ionic current and the precipitation cycle. Lightning and point discharge restore electrons to the earth. The arrows show the movement of electrons in maintaining earth's electrical balance.

The normal atmospheric electric polarity, as shown in Fig. 4-2, is reversed under a thundercloud. As shown in Fig. 4-7, the "sky" or cloud base is strongly negative with respect to the earth. This voltage forces the electrons to flow through nature's points in the reverse direction when a thunderstorm is near. The normal voltage gradient of 150 volts per meter is replaced by an opposite potential of up to 10,000 volts per meter, or greater. The electron flow in Fig. 4-9 maintains a balance between those moving from the earth and those moving toward the earth.

## MEASURING THE ATMOSPHERIC CHARGE

Table 4-1 shows the estimated sheet for the yearly transfer of electrons in coulombs per square kilometer. One coulomb represents the flow of 6,242,000,000,000,000,000 electrons. So, an estimated 120 coulombs flow from earth and to earth for each square kilometer of the earth's surface. Multiplying this by the surface area of the earth gives the total number of electrons flowing from and toward the earth. This becomes a gigantic number.

There are several ways to get an indication of the atmospheric electrical gradient. A serious word of warning. The voltages encountered in the atmosphere are large, and can generate violent and lethal discharges. People have been seriously injured or killed by the atmospheric electricity. One easy

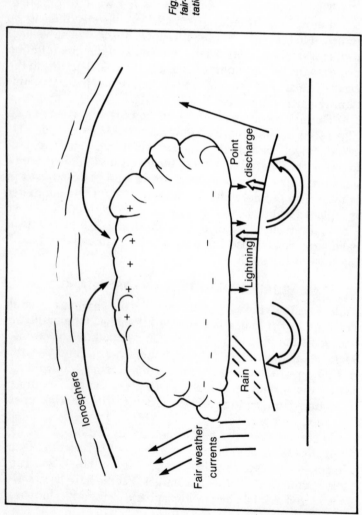

Fig. 4-9. Charge movement cycle for fair-weather ionic current and precipitation cycle.

**Table 4-1. Estimated Balance Sheet Showing Yearly Electricity Transfer For the Entire Globe, Measured in Coulombs Per Square Kilometer.**

|  | From Earth | To Earth |
|---|---|---|
| Fair weather Ionic current Precipitation | 90 30 | 0 0 |
| Point discharge Lightning | 0 0 | 100 20 |

way to get hurt is to try to fly a kite in a thunderstorm, or to fly a kite with a metallic kitestring. This is asking for trouble, especially if it is being held. Remember that this high-voltage gradient exists, even on a clear day. Anything that can conduct electrons from one level to another without dissipating them can carry a lethal charge.

One method of indicating the charge in the atmosphere is to run an uninsulated conductor between two clothesline poles. Insulate the wires from the poles, and run the insulated lines to a highly sensitive meter, such as an electrostatic voltmeter. This will sometimes work with an electroscope (Fig. 4-10). The bare wire runs back and forth four times be-

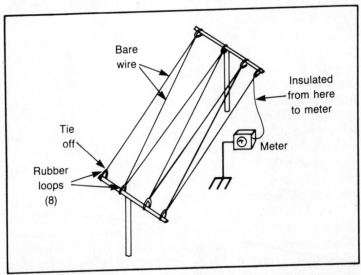

Fig. 4-10. Using a wire strung on clothesline poles to obtain an indication of the atmospheric charge.

tween the poles, held at each end with a rubber loop made from an old inner tube.

The rubber inner tube insulates the wire from the poles, so that the wire is not grounded. The portion of the wire from the clothesline pole to the meter is insulated, so any electrical charge that is picked up will not leak off.

Realize that most clotheslines are 2 meters or less high. With a voltage gradient of 150 volts per meter, this will give a potential of 300 volts. This is the maximum available, and measuring it is difficult because the electrons are in the air, and must be picked up by a conductor. For the clothesline, allow the wire to remain connected to the poles, with the free end not connected to anything for several hours so that the charge builds up on the wire. Then connect the free end of the wire to the indicator. There will be one short discharge, which might be in the form of a small spark. This spark can sometimes be observed using the neon light as described in Chapter 3.

Remember that the charge takes time to build up, and one shot might be all there is without giving it time to recharge. The rate of charge buildup is a function of the surface area exposed to the charge. The more area the faster the charge buildup. Also, this works the best in dry weather with low humidity, so there is less tendency for the charge to bleed off.

B. F. J. Schonland actually measured the current of point discharge from a tree in South Africa in 1927. As illustrated in Fig. 4-11, he insulated a tree and measured the current flowing from the tree to the atmosphere. Because the galvanometer (microammeter) is connected in series between the tree and ground, the electron flow will be through the galvanometer. There will therefore be a reading of the current, which is a measure of the electron flow. This experiment showed that electrons pass from the earth, through the tree, and out into the air during fair weather. The tree leaves and branches make up many point discharge points, so this is a measure of the point discharge of the tree.

Dr. Schonland also found that when thunderclouds were overhead, electrons leave the air, enter points on the tree, and flow to the earth with greater magnitude than the fair weather example. This graphically demonstrated that the current flow reversed when a thundercloud was overhead.

This experiment can be duplicated using a sensitive gal-

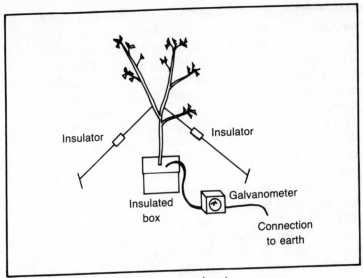

*Fig. 4-11. B. F. J. Schonland's tree experiment.*

vanometer and the proper current shunts. Several things can be used in place of a tree: a plant with several pointed leaves, or even a square of good grass—as long as there are several points pointed skyward. The container must be insulated. A good-quality plastic container resting on some wooden or plastic blocks will suffice. The ground must be a good ground, such as a rod driven into damp soil. The ground improves if the soil is damp—that is, if the ground rod does not corrode. If required, it is best to have the help of someone versed in electronics and electricity to help set up the galvanometer. By keeping a record, at least daily, with additions whenever the current or weather changes, the changes in the point discharge current can be seen. Record the current, direction, temperature, time, cloud conditions, and weather.

## LIGHTNING

Lightning is the sudden release of charges in an effort to balance the charge on the ground with the charge in the clouds. Most lightning occurs as a result of thunderstorms. There are other types of lightning—such as heat lightning—that have their source with hot, dry air and static charge buildup.

Mankind marveled at lightning, and believed it to be supernatural and sacred. The thunderbolt, as hurled by Zeus or flickered in the wink of Thunderbird's eye, marked significant and sacred places. It was the ultimate weapon of the primitive gods. For both religious and practical reasons, mankind has given the thunderbolt a measure of respect.

Today, the study of lightning is more scientific. The techniques of investigation are more experimental. Still, we are awed and frightened when lightning strikes nearby, and we feel the charged air, see the flash, and hear the thunder. But lightning is also dangerous: every year it kills about 150 Americans and injures about 250. This is more than the average annual death toll from tornadoes or hurricanes.

Lightning is also very destructive. The current peaks in a lightning strike can reach 200,000 amps or more. Currents of this magnitude have a crushing effect on conductors, and can cause non-conducting or semiconducting materials like wood or brick to explode violently. The current produces heat, and is responsible for many fires. Any item—a car, airplane, house, or structure—when hit by lightning, will suffer damage. In addition, livestock and wild animals are killed, and electromagnetic transmissions are disrupted. The total estimated damage from lightning is more than $100 million annually.

The earth is normally negatively charged with respect to the atmosphere. As a thunderstorm passes over, the negative charge in the base of the cloud induces a positive charge on the ground below, and several miles around the cloud. This ground charge follows the thunderstorm like an electrical shadow. The ground charge grows stronger as the negative cloud charge increases. The attraction between positive and negative charges forces the positive ground current to flow up buildings, trees, and other elevated objects in a effort to establish a flow of current. But air, which is a poor conductor of electricity, insulates the cloud and ground charges, preventing more than a corona current flow until large electrical charges build up. This is illustrated in Fig. 4-12, which shows a thunderstorm and ground charge.

Lightning occurs when the difference between the earth charge and the cloud charge becomes great enough to overcome the resistance of the insulating air. This forces a conductive path for current to flow between the two charges. The voltage in these cases can be as much as 100 million volts. Lightning strokes usually represent a flow of current from

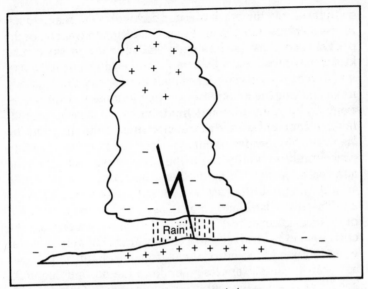

Fig. 4-12. A thunderstorm, showing the ground charge.

negative to positive, i.e., electron flow from the cloud to the earth. Lightning can also proceed from cloud to cloud, from cloud to ground, or, where high structures are involved, from ground to cloud (Fig. 4-13). Notice that there is a small pocket of positive charges in the lower regions of the thunderstorm in this illustration; this may, or may not be present when lightning occurs.

The typical cloud to ground lightning stroke begins as a pilot leader, too faint to be visible. This leader advances downward from the cloud, and establishes the initial portion of the stroke path. Surge currents, called step leaders, follow the initial leader. These step leaders advance the stroke about 100 feet or more at a time, toward the ground. They pause and repeat the sequence until the stroke is near the ground. This establishes a conductive path of electrified (ionized) air from the cloud to near the ground. There, discharge streamers from the ground reach up to intercept the leader path, and the conductive is completed between the ground and cloud charges. When this path is complete, a return stroke leaps upward at speeds approaching the speed of light. This return stroke illuminates the branches of the descending leader track. Because the initial leader track points downward, the stroke appears to come from the cloud. The brightness of the stroke is the result of glowing atoms and molecules of air energized by the stroke.

Once the channel has been established and the return stroke has ended, dart leaders from the cloud initiate secondary returns. This continues until the opposing charges are dissipated, or the channel is gradually broken by air movement. Even when luminous lightning is not visible, current can continue to flow along the electrified channel set up by the initial step leader. What appears to be a single lightning stroke might in reality be several different strokes along the same path, and occurring in a very short time. The total time for a typical lightning stroke is less than one second.

The ground to cloud discharges occur less frequently. Because they consist of discharges in the opposite direction, the initial leader is from the earth to the cloud. The initial leader is not followed by a return stroke from the cloud. This might be because charges are less mobile in the partially conductive clouds than in the highly conductive earth. Once the conductive path is established, however, the current flow may set up sequences of dart leaders and returns.

Fig. 4-13. Several different lightning paths within a thunderstorm.

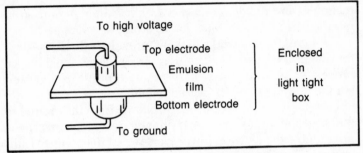

*Fig. 4-14. The Klydonograph for monitoring lightning strikes.*

Thunder is the crash and rumble associated with lightning. It is caused by the very rapid heating of air in the stroke. This rapid heating causes an explosive expansion of the air, which produces the noise. When close to the lightning, it makes a sharp explosive sound. More distant strokes produce the familiar rumble. This is a result of the sound being refracted and modified by the turbulent environment of the thunderstorm. Because light travels so many times faster than sound, the distance from the lightning can be estimated by counting the number of seconds between the lightning and the thunder, and dividing by five. This gives the distance in miles, because sound travels at about one mile in five seconds.

The dual character of lightning, the high currents, and the destructive heat make it doubly dangerous. The high currents, which can reach 200,000 amperes or more, produce exceptional forces that can destroy buildings or trees. The heat produced by the expanding air, and by the conduction of the current, is responsible for numerous fires.

The use of high-speed photography and sensitive camera films have allowed scientists to capture the lightning stroke on film. This has shown both cloud to ground and ground to cloud lightning, and how the lightning stroke propagates.

In 1924, J. Peters, of the Westinghouse Company developed the Klydonograph, to monitor for lightning strikes. The Klydonograph (Fig. 4-14) records images on the film. The shape of the image indicates the polarity, and the size is a function of the magnitude. The Klydonograph has served as a silent watchman on remote locations of transmission lines to give an approximate indication of lightning discharges. Presently other more complex devices, such as surge voltage recorders and remote ground current monitors, are used to monitor transmission lines for lightning discharges.

A lightning rod works in that it provides a discharge path for the lightning stroke. The lightning rod, mounted on the highest point, and being sharp pointed, provides a ground corona discharge. As discussed before, pointed objects tend to be a point of corona current discharge from the ground to the atmosphere. When the thunderstorm passes over, the direction of the electron flow reverses, but the corona is still present. This can discharge the atmosphere rapidly enough that lightning does not occur. If lightning does occur, the corona current ionizes the air around the tip of the lightning rod, providing a partially ionized path for the lightning stroke. The good, solid ground connection to the lightning rod grounds the lightning stroke so that it does no damage.

The effective area of protection provided by a lightning rod has been determined, and the results are available in construction and National Fire Protection Agency handbooks.

A lightning strike near a long conductor, such as a metal fence or power line, can induce a sharp wave of current to move along the conductor away from the lightning strike, as in Fig. 4-15. A lightning strike occurring near a power line

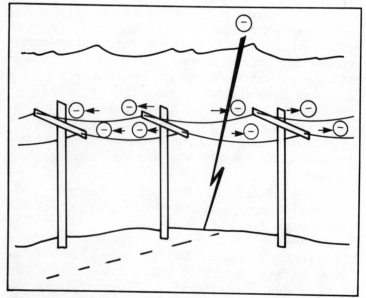

Fig. 4-15. The effect of a lightning stroke near a power line.

contains a large negative charge. This charge repels electrons (negative charges) both directions down the line away from the strike. This sharp wave will move down the conductor until the stroke is over. Then the wave might return to restore conditions back to normal.

## SFERICS

In addition to the liberation of energy by lightning, a thunderstorm produces a form of electromagnetic waves called sferics. These account for the familiar "static" heard on radio receivers when a thunderstorm is in the area. A portion of the energy escapes into space, but most of it is trapped between the earth and the ionosphere until it is dissipated. So, when the electric field pattern of the atmosphere is disrupted, and lightning discharges are present, sferics are radiated.

The field disturbance can be detected and measured over an area of 50 to 200 square miles. A simple method to detect an atmospheric field disturbance is to measure the direction and magnitude of current flow through a lightning rod type of arrangement, as shown in Fig. 4-16. Make sure that the point of the rod is not above the surrounding structures and trees. It must not act as a lightning rod, this can destroy the meter. The polarity of the current indicates if a disturbance is present, and the meter reading gives an indication of the field strength. Plot the polarity and reading against the weather conditions for a period of time. An interesting correlation can be determined.

The sferics can be detected over hundreds of miles. Because they are electromagnetic radiation, they can be detected using a radio receiver. By looking at sferics with an oscilloscope, it has been determined that no two sferics are alike, and that they cover a broad spectrum of frequencies. Most of the energy is concentrated below 20 kilohertz, although it extends as high as 5,100 kHz.

One experimenter reported that by counting the sferics occurring at 430, 2,000 and 5,100 kilohertz, he was able to determine the distance to the storm, and obtain a measurement of the intensity. Typically, the 430 kHz frequency has the most number of sferics, with the 2,000 frequency next and the 5,100 the lowest number. Most of those appearing on the two higher frequencies also appear on the 430 kHz band. This

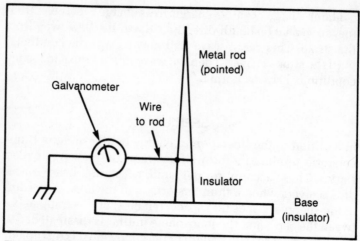

*Fig. 4-16. A simple method to determine the presence of an atmospheric electric field disturbance.*

has been tested by determining the time coincidences between sferics appearing on the different frequencies.

The 430 kilohertz band will provide consistent data up to a radius of 200 miles, with counting rates frequently as high as 150,000 sferics per minute. The range of the higher frequency bands is less, resulting in a lower count.

During thunderstorm activity, there are many lightning discharges and, consequently, many sferics. During clear days the distribution of the electric field between the earth and the ionosphere is relatively uniform. Even during these clear days, sferics can be generated as a result of the corona discharge between the atmosphere and the earth. It should be noted that sferics are associated with electrical discharges, and that they might or might not be visible. The visible discharges are lightning, while those that are not visible are the corona discharges.

The local library has books on natural science and atmospherics that will cover sferics in much greater detail. They also give the details of a detector. To build such a detector will require considerable research and effort, but when it is built, some very interesting and entertaining experiments can be conducted. These include time correlation on all the frequencies, and an attempt to determine the magnitude of the sferic. Also, the occurrence can be correlated to the weather, in both the local area and in the surrounding areas.

## USING ATMOSPHERIC
## ELECTRICITY TO DRIVE A STATIC MOTOR

Because there is an electrical field in the atmosphere, it is logical that this field can be used to drive a static motor. This is possible, and has been demonstrated. One experimenter reported success using a small helium-filled balloon to lift a bare metallic conductor into the air. The electrostatic motor was set up on the roof of a building. One end of the bare conductor was connected to the input to the static motor, and the other end connected to the balloon. Let the balloon float in the atmosphere for an hour or so before connecting the conductor to the motor. This will allow the conductor to charge up from the atmosphere.

A similar experiment was tried with the balloon rising between buildings, and not rising above them. Because the buildings were still at ground potential, there was basically no charge for the conductor to pick up, so the static motor could not be run. Another experiment consisted of running the static motor from a long whip antenna. Although when operated from the rooftop, the whip antenna produced results, it was not as good as the wire supported by the balloon. This is to be expected because the amount of charge induced into the wire is a function of the length of the antenna, and the direction. In both cases the antenna pointed up, but in the first case the antenna was longer.

When experimenting with atmospheric electricity, be careful not to touch the conductor or anyplace in the electron flow path. This could result in a sizable jolt of electricity. Always treat it with respect, because it can be dangerous. Use a ground bar or lead to ground points that might have an electrical charge, before touching them. If the balloon is left airborne for any length of time without some means of depleting the charge, a sizable charge can build up on the wire. Therefore, if the balloon is not to be used for several hours, bring it down. At the very least, short the lower end of the bare wire to ground. This will ground out any built-up charge.

Helium-filled balloons can usually be obtained at carnivals, fairs, parades, and things like that. The bare wire is another problem. If it is possible to obtain a spool of nickle fishing line, this will work. Some sporting goods supply stores will carry it for deep sea fishing.

It is advisable to keep the length short enough so that you don't exceed the Federal Aviation Administration height requirements. This is especially critical if you are in the landing pattern or near an airport.

Make up a simple wooden spool to hold the wire. Figure 4-17 shows one type of such a spool. The wire is wound on the drum lengthwise, and is unwound in the same manner. This helps reduce the chance of the line becoming twisted. It is mounted on a wooden platform that is weighted down. The handle is tied off to lock the drum and prevent letting out more line.

Instead of nickle fishing line, it is possible to obtain some bare copper wire, of a size between 20 and 26. Copper wire can be more flexible, but it is also more apt to break. In either case, handle the wire with care and prevent kinking the wire. A kink results in a weak point that could break.

The line or wire must be fastened securely to the balloon. This can be done using a short length of string tied around the neck of the balloon. This string can be securely tied to the line or wire. In addition, a crimp-on terminal can be attached to the line or wire, and this tied to the balloon.

Another method to pick up the atmospheric charge is to use a bare wire strung between two poles, and insulated from the poles. Again, the wire will pick up an electrical charge from the atmosphere, but this will take time because the air is a gas, and non-conductive. Therefore, the charges in the air must come in contact with the wire. One way to reduce the charging time is by passing a flame along the wire. The flame produces ions in the vicinity of the wire, and in a few seconds the potential of the wire will reach that of the am-

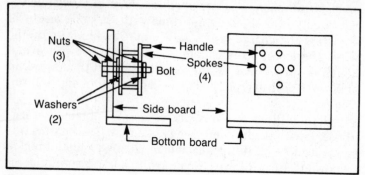

Fig. 4-17. A simple spool to let out a helium balloon.

Fig. 4-18. Lightning (Courtesy of NOAA).

bient air. This flame can be produced by wrapping the wire in paper soaked in a flammable liquid, such as lead nitrate. Then the paper is ignited, and a flame along the wire results. Do not use a liquid that evaporates quickly, such as gasoline or lighter fluid, because it might all evaporate before it can be ignited. It is also dangerous.

This charge might not be enough to drive a monitor continuously, so that when continuous monitoring is desired, some means is employed to generate a continuous supply of ions in the vicinity of the wire. One method is to deposit a thin layer of radioactive material on the wire. This will emit alpha particles that ionize the surrounding air, resulting in a continuous charge. That is, if the monitoring instrument does not draw off the charge faster than it is charged.

There are other atmospheric experiments that can be done. Here is one: String some bare wire between two poles, insulating the poles from the end. Connect it to a neon bulb, and determine how often the bulb fires. The bulb must be mounted on a black box, so that single firings of the bulb can be seen.

Atmospheric electricity is there, and its effects can be observed (Fig. 4-18). The whole field of nature's electricity is a very interesting one to pursue, to learn more about. It is hoped that the reader will delve deeper into these areas of interest.

# Index

# Index

**A**

aluminum, 8
  atomic diagram for, 9
  conduction in, 12
amplifier, 45
  boosting circuit gain by adding, 47
  parts list for, 46
  schematic for, 46
  use of, 47
atmosphere
  circuit equivalent of, 119, 120
  fair weather current flow in, 118
  zones of, 117
atmospheric charge
  determining presence of field dis-
    turbance, 138
  experiments to measure, 140
  global transfer per square kilome-
    ter, 129
  lightning in, 131
  measuring the, 127
  obtaining indication of, 129
  sferics in, 137
  static motors and, 139
atmospheric electricity
  sources of, 122
atom, 5
  diagram for, 7
atomic weight, 7
audio squealer, 36
audio tone generator, 36
aurora borealis, 3

axes, 41

**B**

ball lightning, 65
Bohr's atomic model, 7

**C**

capacitor, 59
charge, 5
  equation for, 42
  generation of, 39
  shear generated, 40
charge direction of, 41
charges
  movement within a solid, 17
compression, 40
conduction, 19
conductor, 17
corona effect, 122
corona motor, 86-94
  comb assembly of, 92
  construction details for, 89
  shaft and rotor assembly of, 90
cosmic rays, 4, 116, 123
crystalline structure, 31
  materials and chemical composi-
    tion of, 32
Curie, Pierre and Jacques, 28
current flow, 6

**D**

d constant, 41

depth finders, 28

**E**

earthquakes, 3, 55
electret, 94-95
electrical apparitions, 3, 55
electrical field disturbance, 138
electricity, 2
  atmospheric, 116
  natural generators of, 4
electromotive potential, 5
electron flow, 5, 118
electrons
  shared, 17, 18
electrophorus, 80, 81
electroscope, 18, 19
  conduction charging of, 21
  details of, 25
  homemade, 23
  plans for a, 24
  twin electrode, 22
electrostatic generator, 77
  details of, 78
electrostatic motors, 86
  corona, 86
electrostatic voltmeter, 66
electrostatics, 58
elements and their atomic properties, 8
energy
  electrical, 2
  nuclear, 2
  potential, 1
  solar, 1
  steam, 2
exosphere, 117

**F**

"Foo Fighters," 65
force
  direction of, 41
Franklin, Benjamin, 58, 108, 121

**G**

g constant, 41
ground, 22, 64
ground charge, 133
Guericke, Otto von, 58

**H**

headphones, 45
high-frequency mechanical vibration, 30
"hole," 14

**I**

induction, 19

inductive pickup, 51
ionic current
  fair weather charge movement cycle of, 128
ionization, 122
ions, 32
  formation of, 33
  movement of, 33
  negative, movement in solid, 34
  positive, movement in solid, 34

**K**

klydonograph, 135

**L**

Leyden jar, 58, 59
  homemade, 60
  parts list for, 61
  plans for, 61
lightning, 3, 110, 116, 131-137, 141
  klydonograph to measure, 135
  paths of, 134
like charges, 15
liquid crystal display (LCD), 48
  test fixture for, 50

**M**

mechanical stress, 42
meter, 45

**N**

natural acid batteries, 4
negative ion, 13
non-conductor, 17
northern lights, 3

**O**

opposite charges, 16
orbit, 9
oscilloscope, 44

**P**

periodic table (partial), 8
Peters, J., 135
phonograph cartridge, 36
piezoelectric ignition device, 38
piezoelectricity, 28
  applications of, 35
  changing size of, 29
  characteristics of, 39
  crystalline structure and, 31
  d33, g33, and k33 values of, 44
  demonstrations of, 43
  effect of time on maximum, 52
  generating and moving ions within, 35

ions and, 33
liquid crystal display and, 48
waveforms for 1000-hour pressure
application, 53
pith ball, 69
plasma ball, 65
Poggendorff, J.C., 87
point discharge, 122
polarity, 13, 15
positive ion, 14
potential gradient, 119, 121

**Q**

quartz crystal, 32

**R**

Rochelle salts, 53

**S**

salt water electrolysis, 4
Schonland, B.F.J., 130
Schonland's tree experiment, 131
sferics, 137-139
shear, 39, 40, 41
sodium chloride
atomic structure of, 30
solar wind, 122
solid matter, 30
crystalline, 31
noncrystalline, 31
sound
definition of, 37
generation and propagation of, 38
St. Elmo's fire, 122
static charge, 26, 62
plastic buildup of, 60
static electricity, 3, 16, 116
air generation of, 64
demonstrations of, 72
detection of, 67, 69
electret and, 94
electrophorus and, 80
electrostatic motors and, 86
generation of, 73-77
history of, 57
Leyden jar and, 58
low and high humidity voltages
for, 66
neon lamp to detect, 68
triboelectric series, 61
Van de Graaff generator and, 96
versorium, 84

voltages encountered in, 65
voltmeter for, 66
static electrictiy
generators of, 77
static motors
using atmospheric electricity to
drive, 139
stratosphere, 117

**T**

tension, 40, 41
thermoelectricity, 5
thermosphere, 117
thunderstorms, 116
profile of 124-127
triboelectric series, 61, 63
troposphere, 117

**U**

ultrasonic cleaners, 28

**V**

Van Allen belts, 116, 123, 124
Van Allen, James, 123
Van de Graaf, Robert J., 96
Van de Graaff generator, 58, 80,
96-115
aurora borealis effect with, 107
ground terminal for, 109
insulated table for, 109
lower portion of, 101
metal screen electrodes for, 112
needle assembly for, 106
neon lamp test probe for, 105
overall view of, 98
schematic of, 97
simulated lighning with, 110
test probe for, 104
upper terminal assembly of, 100,
103
upper tube end for, 114
Varley generator, 80
versorium, 84
voltage, 119
changing piezoelectric size with,
29
equation for, 43
piezoelectric generation of, 29
static electricity, 65
voltage gradients
linear and non-linear, 120